walkernaths 1.4

LINEAR ALGEBRA

NCEA Level 1 Internal

Charlotte Walker and Victoria Walker

Australia • Brazil • Japan • Korea • Mexico • Singapore • Spain • United Kingdom • United States

Walker Maths 1.4 Linear Algebra
1st Edition
Charlotte Walker
Victoria Walker

Editor: Eva Chan
Designer: Cheryl Smith, Macarn Design
Production controller: Siew Han Ong

Any URLs contained in this publication were checked for currency during the production process. Note, however, that the publisher cannot vouch for the ongoing currency of URLs.

Acknowledgements
Cover photo courtesy of Shutterstock.

For product information and technology assistance,
in Australia call **1300 790 853**;
in New Zealand call **0800 449 725**

For permission to use material from this text or product, please email **aust.permissions@cengage.com**

National Library of New Zealand Cataloguing-in-Publication Data
A catalogue record for this book is available from the National Library of New Zealand.

978 0 17 037043 1

Cengage Learning Australia
Level 7, 80 Dorcas Street
South Melbourne, Victoria Australia 3205

Cengage Learning New Zealand
Unit 4B Rosedale Office Park
331 Rosedale Road, Albany, North Shore 0632, NZ

For learning solutions, visit **cengage.co.nz**

Printed in China by 1010 Printing International Limited
13 14 15 16 17 27 26 25 24 23

CONTENTS

ISBN: 9780170370431

Glossary

Make your own glossary of key terms:

Term	Definition	Picture/Example
Linear		
Gradient		
Intercept		
Discrete data		
Continuous data		
Variable		
Constant		
Coefficient		
Origin		

ISBN: 9780170370431

Substitution

- When substituting, you replace **variables** with **numbers**.
- It is important to remember **BEDMAS** when doing this.

Examples: If $w = 4$, $x = 8$, $y = -2$ and $z = -1$, find the value of A for the following.

1 $A = w(x - 4z) - y$
$= 4(8 - 4(-1)) - (-2)$
$= 4(12) + 2$
$= 50$

2 $A = w - \frac{x - y}{2(z - y)}$
$= 4 - \frac{8 - (-2)}{2((-1) - (-2))}$
$= 4 - \frac{10}{2}$
$= 4 - 5$
$= -1$

If $b = 6$, $c = 3$, $d = -4$ and $e = -2$, calculate the values of A.

1 $A = 7d + 13$

2 $A = 2d + 3c$

3 $A = c(b - e)$

4 $A = e(b - c) - d$

5 $A = c + e(b - d)$

6 $A = 2b - e(d - 4c)$

7 $A = \frac{b}{2}(b - c)$

8 $A = \frac{b - d}{e}$

9 $A = c(b + d) + (d - e)$

10 $A = \frac{b + d}{c + e}$

11 $A = \frac{d(b - c)}{cd}$

12 $A = b - \frac{d - e}{2(c + e)}$

ISBN: 9780170370431

Solving linear equations

'Solve' means 'find a value for x'.

Rules: 1 You can do anything you like to an equation as long as you do the **same to both sides**.
2 There should be only **one equals sign** per line.
3 Collect all the variables on one side and numbers on the other side.
4 When you want to get rid of something, perform the **opposite** operation.
5 Your answer should always be in the form **x = ...**

Trick: If you need to change the sign of everything, multiply **both** sides by **-1**.

One-step equations

Examples:

1 $x + 12.3 = 10$ **(– 12.3)**
$x = -2.3$

2 $6.3 - x = 14.9$ **(– 6.3)**
$-x = 8.6$ **(x by -1)**
$x = -8.6$

3 $-5x = 125$ **(÷ by -5)**
$x = -25$

4 $\frac{x}{5} = -3.2$ **(x by 5)**
$x = -16$

Solve the following.

1 $x - 4.6 = 17$

2 $x + 4.8 = 7$

3 $5x = 17$

4 $\frac{x}{4} = 1.25$

5 $x - 29 = -117$

6 $0.82 + x = -1.36$

7 $11.2 - x = 3.9$

8 $-2\frac{1}{2}x = -1.75$

9 $0.9 = -x + 4.1$

10 $-8.01 = -2.73 - x$

 ISBN: 9780170370431

Two-step equations

Tricks: 1 Do the adding or subtracting **before** the multiplying or dividing.

2 If the variable (e.g. x) is on the right, **swap** the sides using the '=' sign as a centre.

e.g. $3.5 = 6x + 1.7 \Rightarrow 6x + 1.7 = 3.5$

Examples:

1

$3.5 = 6x + 1.7$

$6x + 1.7 = 3.5$ **(– 1.7)**

$6x = 1.8$ **(÷ by 6)**

$x = 0.3$

2

$1 - 1.5x = -12.5$ **(– 1)**

$-1.5x = -13.5$ **(÷ by -1.5)**

$x = 9$

Solve the following.

1 $5x - 7 = 39$

2 $0.8x + 2.8 = 10$

3 $17 + 1.6x = 28.2$

4 $199 - 12x = 7$

5 $0.04x - 3.8 = -4.2$

6 $110x + 9.7 = 4.2$

7 $125x - 75 = 0$

8 $213 = 13x - 138$

9 $-7.4 = 0.5x - 16.7$

10 $-0.013 = 0.029 + 3x$

ISBN: 9780170370431

Equations with a variable on both sides

Remember — collect all the terms that include an 'x' on one side and the constants on the other.

Examples:

1
$5 - 22x = 77 - 6x$ **(+ 6x)**
$5 - 16x = 77$ **(– 5)**
$-16x = 72$ **(÷ by -16)**
$x = -4.5$

2
$20x - 1.8 = 7.3 + 7x$ **(– 7x)**
$13x - 1.8 = 7.3$ **(+ 1.8)**
$13x = 9.1$ **(÷ by 13)**
$x = 0.7$

Solve the following.

1 $7x - 4 = 4x + 11$

2 $215 - 4x = 15x + 63$

3 $7x - 19 = 18x - 90.5$

4 $1 - 5x = 2x + 102.5$

5 $2.5x - 408 = 26.5x$

6 $26x - 34.6 = 8x + 41$

7 $1 - 29x = -139 - 21x$

8 $3 - x = 9x + 536$

9 $2x + 176 = 11x + 5$

10 $12x + 0.4 = -0.2 + 5x$

ISBN: 9780170370431

Equations with brackets

Expand the brackets, then solve as you did in the last exercise.

Examples:

1 $4(3x - 1) = 5(2 + x)$
$12x - 4 = 10 + 5x$ **(– 5x)**
$7x - 4 = 10$ **(+ 4)**
$7x = 14$ **(÷ by 7)**
$x = 2$

2 $9(3 - x) = 4(7 - x) - 2(x + 1)$
$27 - 9x = 28 - 4x - 2x - 2$
$27 - 9x = 26 - 6x$ **(+ 6x)**
$27 - 3x = 26$ **(– 27)**
$-3x = -1$ **(÷ by -3)**
$x = \frac{-1}{-3} = \frac{1}{3}$

Solve the following.

1 $2(3x - 5) = 4(6 + x)$

2 $6(4 - x) = -3(7x + 2)$

3 $11(x - 2) = -5(3 - 2x)$

4 $6(3 - 4x) = 9(2x + 5) - 6$

5 $8(2x - 3) = 7(5 - 6x) - 1$

6 $6(5x - 4) = 3(8 - x) + 18$

7 $9 - 7(2x - 3) = -2(9 + x)$

8 $4(3x - 1) = 2(7 - 6x) - (18 - 11x)$

9 $-1(5 - 2x) = 6(x - 1) - 3(3x + 2)$

10 $7(2x - 4) - (5 - x) = 3(9 - 11x)$

ISBN: 9780170370431

Equations with fractions

These become straightforward if you multiply **every term** by the lowest common multiple of the denominators.

Multiply **every** term by 24 because 6 x 4 = 24.

Multiply **every** term by 20 because 4 x 5 = 20.

Examples: 1

$$\frac{5x}{6} - 1 = \frac{3x}{4}$$

$$\frac{5x}{6} \times \frac{24}{1} - 1 \times 24 = \frac{3x}{4} \times \frac{24}{1}$$

$20x - 24 = 18x$ **(+ 24)**

$20x = 18x + 24$ **(– 18x)**

$2x = 24$ **(÷ by 2)**

$x = 12$

2

$$\frac{3x-5}{4} = \frac{4x-1}{5}$$

$$\frac{3x-5}{4} \times \frac{20}{1} = \frac{4x-1}{5} \times \frac{20}{1}$$

$5(3x - 5) = 4(4x - 1)$

$15x - 25 = 16x - 4$ **(– 16x)**

$-x - 25 = -4$ **(+ 25)**

$-x = 21$ **(x by -1)**

$x = -21$

Solve the following.

1 $\frac{2x}{3} + 5 = 13$

2 $\frac{3x}{5} + \frac{x}{2} = 11$

3 $\frac{3x}{4} - \frac{x}{5} = 2.2$

4 $\frac{7x-3}{2} = 7.25$

5 $\frac{2-3x}{4} = 5$

6 $\frac{1+3x}{5} = \frac{5x+2}{6}$

7 $\frac{4-x}{3} = \frac{6-3x}{5}$

8 $\frac{3x-4}{5} - \frac{6-x}{3} = 0$

ISBN: 9780170370431

Forming and solving linear equations

- **Define your variable** using the first letter of the word whose value you need to find.
- If you need to find several values, it is usually easiest to select the **smaller** value as your variable.
- Often it is useful to draw a **diagram**.

Words used for the four basic operations:

+	plus, total, more, and, add(ed), increased by, at least, greater than	–	subtract(ed), less, decreased by, smaller than
x	of, times, multiplied by	÷	divided by, shared between

- 'Double' and 'twice' both mean multiply by 2.

Example one: Nine people go to the zoo. Adult tickets cost \$25 and children's tickets cost \$12. The total cost was \$147. Write an equation and use it to help you find the numbers of adults and children who went to the zoo.

Let **a** represent the number of adults.

Always define your **variable**.

Then

$$25a + 12(9 - a) = 147$$

 # children = 9 – # adults

$$25a + 108 - 12a = 147$$
$$13a + 108 = 147$$
$$13a = 39$$
$$a = 3$$

∴ Three adults and six children went to the zoo.

Always answer the question in a **sentence**.

Example two: Isobel is 14 years old, and she was born when her mum was 26. When will her mum be double Isobel's age? You must write an equation and use it to solve this problem.

Let **y** be the number of years until Mum is double Isobel's age.

Variable defined.

Current ages: Isobel is 14 → $+ y$ years → $14 + y$; Mum is 40 → $+ y$ years → $40 + y$

Mum is double Isobel's age when:

$$2(14 + y) = 40 + y$$
$$28 + 2y = 40 + y$$
$$y = 12$$

Answer in a **sentence**.

∴ In 12 years' time, her mum will be 52 and Isobel will be 26, so her mum will be double Isobel's age.

ISBN: 9780170370431

Example three: One side of a triangle is five centimetres shorter than the base. The other side is one centimetre shorter than the base. The perimeter of the triangle is 39 centimetres. Write an equation and use it to calculate the dimensions of the triangle.

Let b equal the length of the base. — **Variable** defined.

Draw a picture:

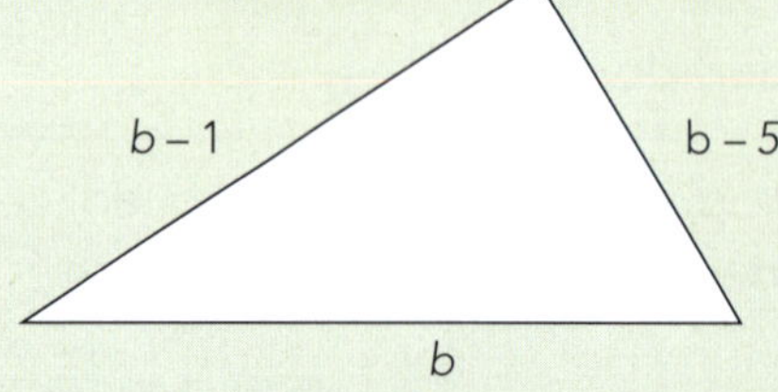

Write an equation:

$$b + (b - 5) + (b - 1) = 39$$
$$3b - 6 = 39$$
$$3b = 45$$
$$b = 15 \text{ cm}$$

Answer in a **sentence**.

∴ The base is 15 cm and the other two sides are 14 cm and 10 cm.

Write and solve equations for each of the following situations.

1 Eight people go to a movie. Adult tickets cost \$15 and children's tickets cost \$8. The total cost was \$85. Write an equation and use it to help you find the numbers of adults and children who went to the movie.

2 Three teachers and 16 students buy tickets for a netball game. Each teacher's ticket costs \$4 more than each student's ticket. The total cost was \$145. Write an equation and use it to help you find the prices for teachers and students who went to the game.

3 Miranda bought seven large and five small hooks from the hardware shop. The total cost was \$58.50. The large ones cost \$1.50 more than the small ones. Write an equation and use it to help you find the price for each type of hook.

ISBN: 9780170370431

4 Henare bought nine trees for his garden. Some were kowhai and some were lancewoods. They cost a total of \$282.50. The kowhai cost \$34.50 each and the lancewoods cost \$27.50 each. Write an equation and use it to help you find the numbers of kowhai and lancewoods that he bought.

5 Siale is five years old, and she was born when her mum was 22. How old will Siale be when her mum is three times her age? You must write an equation and use it to solve this problem.

6 Currently Alfie is 18 years old, and his father is 42. How old was Alfie when his father was three times his age? You must write an equation and use it to solve this problem.

7 Currently Albert is three times Simon's age. After 16 years, Albert's age will be double Simon's age. Calculate their current ages. You must write an equation and use it to solve this problem.

8 Two sides of a rectangle are 3 cm longer than the other sides. Its perimeter is 42 cm. Write an equation relating the perimeter to the lengths of the sides, and use it to calculate the dimensions of the rectangle.

ISBN: 9780170370431

9 Two sides of a parallelogram are 7 cm shorter than the other sides. Its perimeter is 46 cm. Write an equation relating the perimeter to the lengths of the sides, and use it to calculate the dimensions of the parallelogram.

10 One side of a triangle is 23 cm longer than the shortest side. The third side is triple the shortest side. Its perimeter is 83 cm. Write an equation relating the perimeter to the lengths of the sides, and use it to calculate the dimensions of the triangle.

11 Meg, Alannah and Suzie went out for dinner. Meg's meal cost one and a half times Alannah's meal. Suzie's meal cost $3 more than Alannah's. The total cost of their dinners was $66. Write an equation for the total cost of the dinner in terms of *A*, the cost of Alannah's meal. Calculate the cost of each meal.

12 An airline specifies that the maximum sum of the linear dimensions for cabin baggage (length + depth + height) is 118 cm. A bag is two thirds as deep as it is long, and it is 38 cm high. Write an equation relating the dimensions of the bag to the total maximum dimensions. Use it to find the maximum dimensions for this bag.

height
depth
length

13 Paul and Hine have a bag of 18 cherries. Paul keeps most for himself, but gives some to Hine. She complained, so he gave her four more. He then had twice the number of cherries that Hine had. Write an equation and solve it to find how many cherries he gave to Hine at the start.

ISBN: 9780170370431

Rearrangement of expressions

- Expand brackets.
- Put **all** terms that contain the required subject on the left, and everything else on the right.
- Get rid of fractions by multiplying by the denominator.
- Use normal equation-solving rules to isolate the subject.

Examples:

1 Make x the subject of $y - 4 = 3x + 2$.

$$y - 4 = 3x + 2 \quad \textbf{(– 2)}$$
$$y - 6 = 3x \quad \textbf{(÷ 3)}$$
$$x = \frac{y - 6}{3}$$

2 Make y the subject of $y - 4 = \frac{1}{2}(x + 3)$.

$$y - 4 = \frac{1}{2}(x + 3)$$
$$y - 4 = \frac{1}{2}x + 1\frac{1}{2} \quad \textbf{(+ 4)}$$
$$y = \frac{1}{2}x + 5\frac{1}{2}$$

3 Make z the subject of $y + 1 = \frac{5z - 7}{3}$.

$$y + 1 = \frac{5z - 7}{3} \quad \textbf{(x by 3)}$$
$$y\left(\times \frac{3}{1}\right) + 1\left(\times \frac{3}{1}\right) = \frac{5z - 7}{3}\left(\times \frac{3}{1}\right)$$
$$3y + 3 = 5z - 7 \quad \textbf{(+ 7)}$$
$$3y + 10 = 5z \quad \textbf{(÷ by 5)}$$
$$z = \frac{3y + 10}{5}$$

4 Make h the subject of $V = \pi r^2 h$.

$$V = \pi r^2 h \quad \textbf{(÷ by } \pi r^2\textbf{)}$$
$$h = \frac{V}{\pi r^2}$$

Rearrange the following equations.

1 Make x the subject of $y - 7 = 5x + 1$.

____

____

____

____

2 Make x the subject of $y + 4 = 3x - 2$.

____

____

____

____

3 Make x the subject of $3y - 7 = 2x + 5$.

____

____

____

____

4 Make x the subject of $y - 1 = 3(2x + 8)$.

____

____

____

____

ISBN: 9780170370431

5 Make x the subject of $y + 6 = \frac{1}{3}(x - 1)$.

6 Make x the subject of $y + 6 = \frac{3}{4}(x - 2)$.

7 Make x the subject of $y - 5 = \frac{2x - 1}{6}$

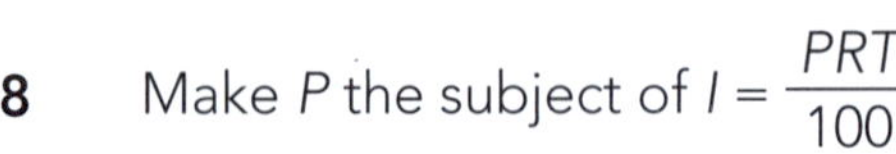

8 Make P the subject of $I = \frac{PRT}{100}$.

9 Make t the subject of $s = v + at$.

10 Make p the subject of $3t(p + q) = a$.

11 The area of a triangle is given by $A = \frac{1}{2}bh$. Write this formula with h as the subject.

12 The sum of the areas of two triangles that have the same height is given by $A = \frac{1}{2}h(B + b)$. Write this formula with B as the subject.

13 Make h the subject of $V = \frac{1}{3}r^2h$.

14 The formula for converting degrees Fahrenheit into degrees Celsius is $F = 32 + \frac{9C}{5}$. Make C the subject.

 ISBN: 9780170370431

Simultaneous equations

- 'Simultaneous' means '**at the same time**'.
- For simultaneous equations, you need to be able to solve two equations that are both true 'at the same time'.
- Because there are **two equations**, there are also **two variables**.
- There are two methods for solving these, and which one you use depends on how the equations are structured.

1 Substitution

- Substitution is easiest where one of the equations is expressed as **x = ...** or **y = ...**.
- As the name suggests, you simply substitute the **x = ...** into the other equation.
- You should number each equation, and say what you are doing at each step.

Examples:

1 Solve the equations $x = 14 - 2y$ and $3y + 4x = 31$.

Number the equations.

$\mathbf{x = 14 - 2y}$ ①

$3y + 4x = 31$ ②

Substitute ① into ②: $3y + 4(14 - 2y) = 31$

From equation ① we know that **x** and **14 – 2y** are equal.

$3y + 56 - 8y = 31$

$-5y + 56 = 31$

$\mathbf{y = 5}$

Substitute for y in ①: $x = 14 - 2(5)$

$\mathbf{x = 4}$ ∴ Solution is (4, 5)

2 Solve the equations $x + 3y = 1$ and $y = 3 - x$.

$x + 3y = 1$ ①

$\mathbf{y = 3 - x}$ ②

Substitute ② into ①: $x + 3(3 - x) = 1$

Select ② because it has the form $y = \ldots$

$x + 9 - 3x = 1$

$-2x + 9 = 1$

$-2x = -8$

$\mathbf{x = 4}$

Substitute for x in ②: $y = 3 - 4$

Select whichever equation is easier for substitution.

$\mathbf{y = -1}$ ∴ Solution is (4, -1)

Solve the following simultaneous equations.

1
$$x = 2y$$
$$4x + y = 9$$

2
$$y = x - 1$$
$$5x - 3y = 9$$

3
$$y = -3x$$
$$x + y = 6$$

4
$$2x - 3y = 15$$
$$x = 2y + 7$$

5
$$y = 15 - x$$
$$3x + 2y = 37$$

6
$$4x + 3y = 26$$
$$y = 12 - 3x$$

7
$$x = 6 - 2y$$
$$x - 2y = 10$$

8
$$x = 15 - 5y$$
$$2x - 3y = 4$$

9
$$x = 2y - 3$$
$$x + 4y - 3 = 0$$

10
$$2x - y - 15 = 0$$
$$y = 6 - x$$

ISBN: 9780170370431

2 Elimination

- Elimination is easiest when the two equations have the **same structure**.

For example:

x terms		y terms	= signs	constants
$3x$	+	y	=	11
$2x$	+	$3y$	=	12

- You solve these by multiplying one or both equations by a constant in order to make either the *x* terms or the *y* terms the **same size** but with **different signs**.
- Then **add** the two equations to produce an equation with one variable only.
- You should number each equation, and say what you are doing at each step.

Examples:

1 Solve the equations $3x + y = 11$ and $2x + 3y = 12$.

	$3x + y = 11$	①
	$2x + 3y = 12$	②
Multiply ① by -3:	$-9x - 3y = -33$	③

Call this ③ because it is a new equation.

Add ② and ③:	$-7x = -21$	
	$\therefore$ **$x = 3$**	
Substitute for x in ①:	$3(3) + y = 11$	
	$\therefore$ **$y = 2$**	$\therefore$ Solution is (3, 2)

2 Solve the equations $4x + 3y = 43$ and $2x + 5y = 53$.

	$4x + 3y = 43$	①
	$2x + 5y = 53$	②
Multiply ① by -5	$-20x - 15y = -215$	③
Multiply ② by 3:	$6x + 15y = 159$	④

You need to multiply **both** equations by a constant.

Add ③ and ④:	$-14x = -56$	
	$\therefore$ **$x = 4$**	
Substitute for x in ②:	$2(4) + 5y = 53$	
	$5y = 45$	
	$\therefore$ **$y = 9$**	$\therefore$ Solution is (4, 9)

Solve the following simultaneous equations.

1
$$2x + 3y = 14$$
$$5x + 3y = 26$$

2
$$5x + 3y = 24$$
$$4x - y = 26$$

3

$$x - y = -7$$
$$4x + 3y = 14$$

4

$$2x + 3y = 19$$
$$4x - y = 3$$

5

$$x + 2y = 15$$
$$5x + 3y = 19$$

6

$$3y + 2x = 26$$
$$4y + 3x = 36$$

7

$$4x - 3y = 9$$
$$3x - 2y = 8$$

8

$$3x - 4y = 20$$
$$4x + 2y = 12$$

9

$$2x + 5y = 19$$
$$5x - 2y = 33$$

10

$$3x - 4y = 14$$
$$2x - 5y = 14$$

11

$$4x + 3y = 43$$
$$2x + 5y = 53$$

12

$$4x - 3y = 22$$
$$3x - 4y = 20$$

ISBN: 9780170370431

Forming and solving simultaneous equations

Hint: Read what the question asks for. Define your variables using their first letters.

Examples:

1 Hugh had three times as many marbles as Kristin. If Kristin had another 20 marbles, she would have had twice as many as Hugh. How many marbles did each person have?

Call the number of marbles Hugh has ***h***.
Call the number of marbles Kristin has ***k***.

Always define your **variables**.

$h = 3k$ ①

$k + 20 = 2h$ ②

Substitute ① into ②: $k + 20 = 2(3k)$

$k + 20 = 6k$ **(– *k*)**

$20 = 5k$

$\mathbf{k = 4}$

Substitute for k in ①: $h = 3(4)$

$\mathbf{h = 12}$

∴ Hugh had 12 marbles and Kristin had 4.

Always answer the question in a **sentence**.

2 Emma buys six apricots and five pears and they cost her \$4.80. Four apricots and three pears cost Erin \$3.10. Calculate the prices of apricots and pears.

Call the price of apricots ***a***.
Call the price of pears ***p***.

$6a + 5p = 4.80$ ①

$4a + 3p = 3.10$ ②

Multiply ① by -3: $-18a - 15p = -14.40$ ③

Multiply ② by 5: $20a + 15p = 15.50$ ④

Add ③ and ④: $2a = 1.10$

$\mathbf{a = 0.55}$

Substitute for a in ①: $6(0.55) + 5p = 4.80$

$5p = 4.80 - 3.30$

$5p = 1.50$

$\mathbf{p = 0.30}$

∴ Apricots cost \$0.55 and pears cost \$0.30.

Write and solve simultaneous equations to find the unknown numbers.

1 Jeggings cost \$18 less than two pairs of shorts. Three pairs of shorts and two pairs of jeggings cost \$160. Calculate the prices of shorts and jeggings.

ISBN: 9780170370431

2 A juicy costs $1.75 less than two ice blocks. Three ice blocks and four juicies cost $6.20. Calculate the prices of ice blocks and juicies.

3 There are three times as many boys as girls in Alexander's class of 28 students. How many boys and girls are there in his class?

4 The number of Merit grades that Rua has is one fewer than double the number of his Achieved grades. If he doubled his Merit grades and added nine, he would have five times the number of his Achieved grades. How many of each grade does he hold at present?

5 Adam has three times as many marbles as Chloe. If Chloe had 80 more, she would have twice as many as Adam. How many marbles does each of them have?

6 Jane bought three pens and four exercise books for $7.95. Sam bought five pens and two exercise books for $6.95. Calculate the prices of pens and exercise books.

ISBN: 9780170370431

7 A kowhai tree costs four times the price of a lancewood. Five kowhai trees and 11 lancewoods cost \$217. Calculate the prices of kowhai trees and lancewoods.

8 Five peaches and three apples cost \$5.10. Two peaches and four apples cost \$3.30. Calculate the prices of peaches and apples.

9 Five pies cost the same as three burgers. Five burgers and two pies cost \$10.85. Calculate the prices of burgers and pies.

10 A T-shirt costs \$15 less than two singlets. Seven T-shirts and four singlets cost \$345. Calculate the prices of T-shirts and singlets.

11 Moana sold packets of seeds to help raise funds for a sports trip. Each packet earned her \$0.65. Afa sold calendars, which earned him \$1.50 each. Altogether they earned \$84.70. If they sold a total of 95 items between them, how many did each student sell?

ISBN: 9780170370431

Straight lines

Co-ordinates revision

Remember: Co-ordinates are in alphabetical order, (*x*, *y*).

Write down the co-ordinates of the lettered points shown at right. The first one is done for you.

a	(2,6)	**b**	(,)
c	(,)	**d**	(,)
e	(,)	**f**	(,)
g	(,)	**h**	(,)
i	(,)	**j**	(,)
k	(,)	**l**	(,)
m	(,)	**n**	(,)
o	(,)	**p**	(,)

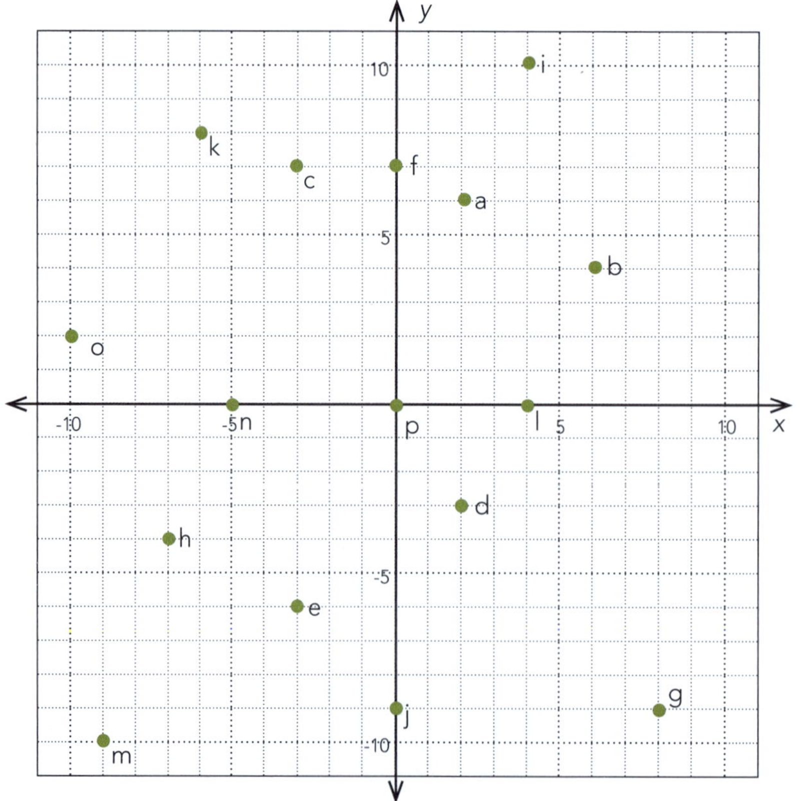

 ISBN: 9780170370431

Plot and label the following co-ordinate points on the grid.

a	(3, 6)	**b**	(-5, 7)
c	(-9, 5)	**d**	(3, -2)
e	(1, -7)	**f**	(-9, -2)
g	(-8, -10)	**h**	(6, -9)
i	(10, 4)	**j**	(-2, 9)
k	(7, 0)	**l**	(-3, -2)
m	(10, -7)	**n**	(-1, 2)
o	(0, -2)	**p**	(-8, 1)

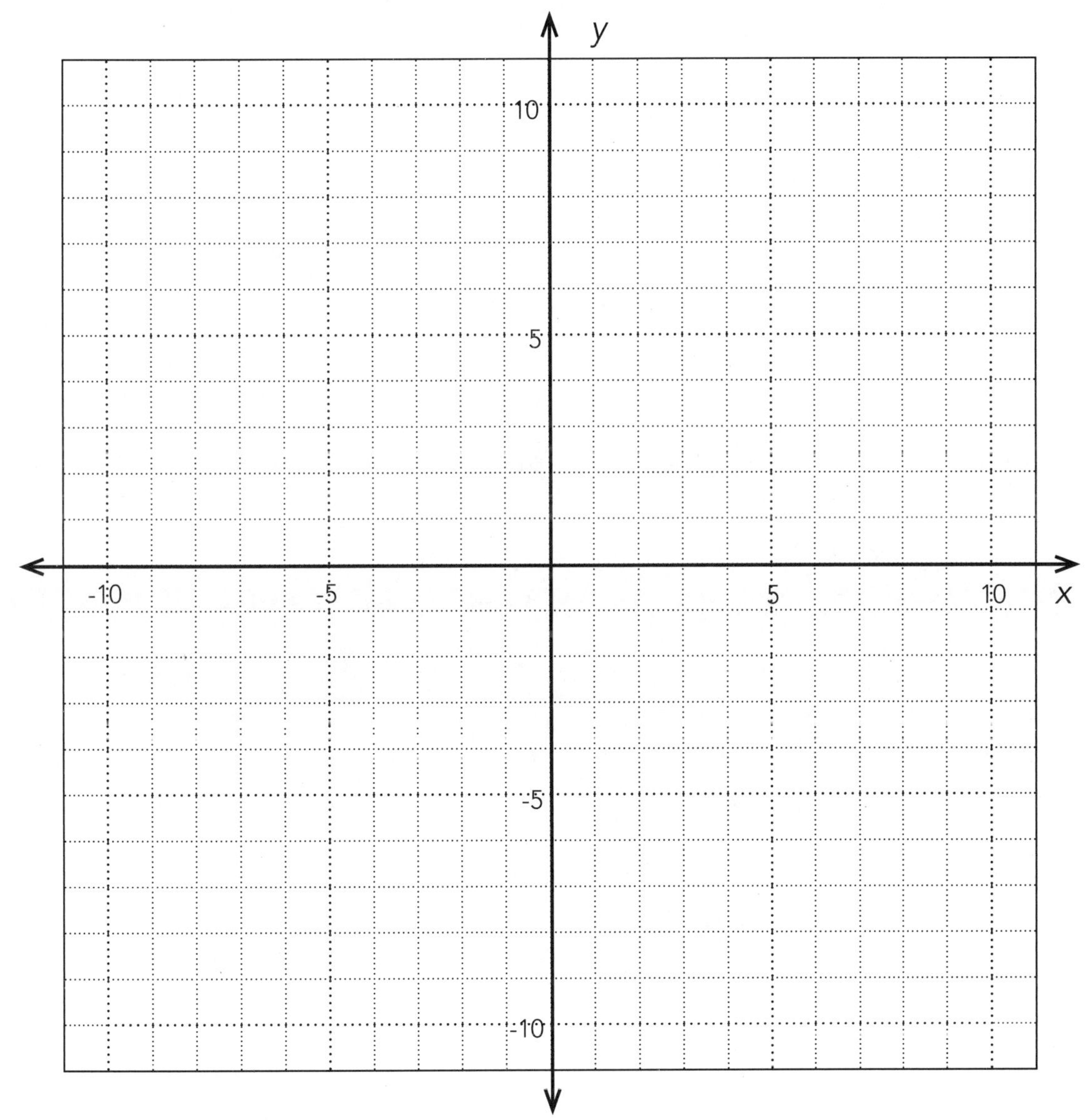

ISBN: 9780170370431

Linear patterns with discrete data

- **Discrete** data is data that can be **counted**, e.g. 'number of …', prices of items that can be bought with only $5 notes, heights that are measured to the nearest centimetre.
- You need to be able to continue a pattern, plot these on a graph, find a rule and use it.

Pattern 1 Pattern 2 Pattern 3

Example:
Alex makes patterns with buttons as shown.

a Complete the table.

Pattern # (n)	# of buttons (B)
1	6
2	8
3	10
4	**12**
5	**14**
6	**16**

+2 +2 +2 +2 +2

Look for the pattern: in this case, **+2**.

b Plot the points on the graph.

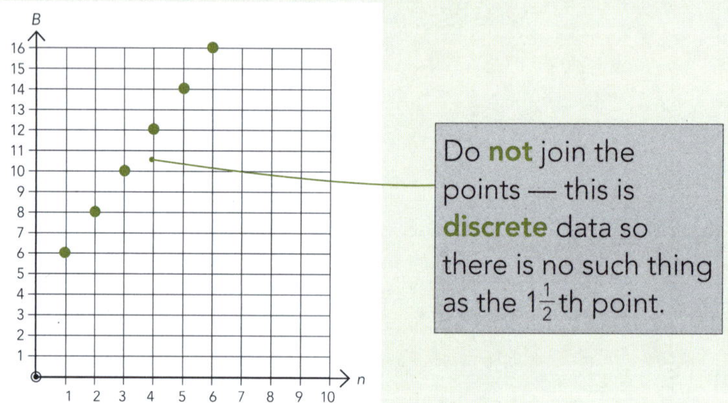

c Give the equation for the number of buttons (B) in any pattern (n).

Use the format $B =$ **+2** $n +$ **+4** So the equation is $\mathbf{B = 2n + 4}$

Insert the **+2** from the table. Insert the # of buttons for pattern number 0.

d Use your equation to find how many buttons would be in his 20th pattern.

20th pattern $\Rightarrow n = 20 \therefore \mathbf{B = 2} \times 20 \mathbf{+ 4} = 44$. So the 20th pattern would have 44 buttons.

e Explain how the equation relates to the pattern.

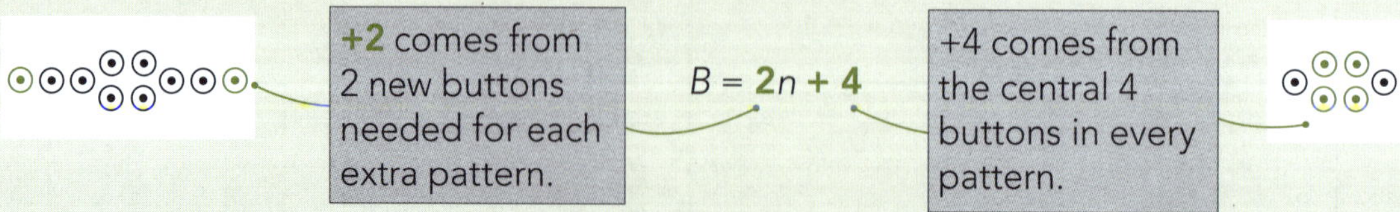

f Which pattern would have 92 buttons?

$92 = 2n + 4$ **(– 4)**
$88 = 2n$ **(÷ 2)**
$\therefore$ $n = 44$. So pattern 44 has 92 buttons.

ISBN: 9780170370431

Answer the following questions.

1 Bridget makes the following pattern with buttons.

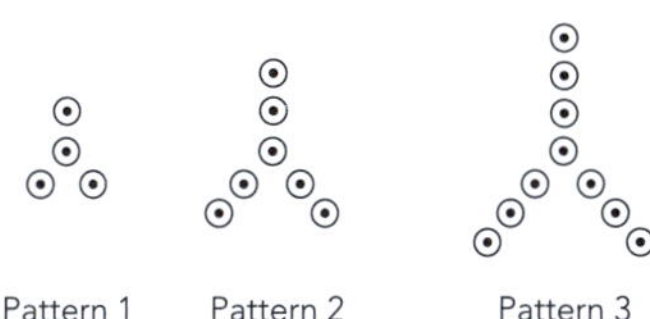

a Complete the table, and use it to find the equation.

Pattern # (n)	# of buttons (B)
1	4
2	7
3	10
4	
5	
6	

Equation: $B =$ ______ $n +$ ______

b Plot the points on the graph.

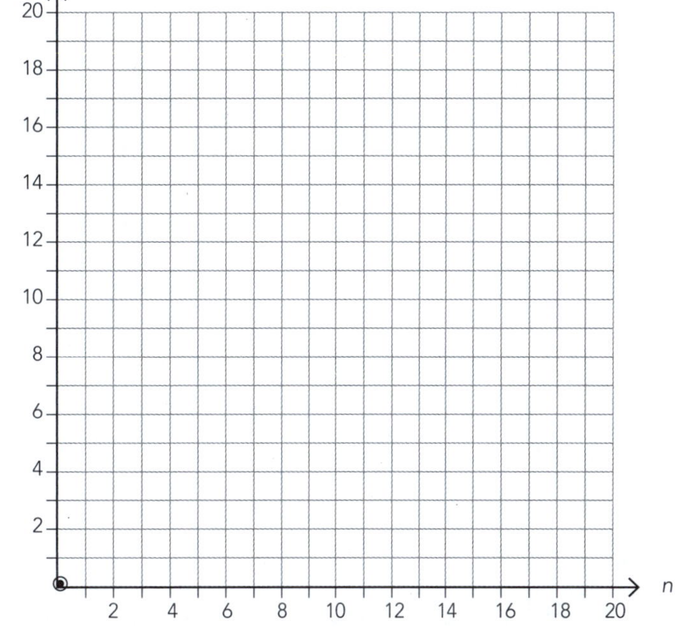

c How many buttons are needed for the 30th pattern?

30th pattern $\Rightarrow n = 30 \therefore B =$ ______ x 30 + ______ = ______.

So the 30th pattern needs ______ buttons.

d Which pattern would have 271 buttons?

271 buttons $\Rightarrow 271 =$ ______ $n +$ ______

____

So the ______ pattern would need 271 buttons.

e How does the equation relate to the pattern?

$B =$ ______ $n +$ ______

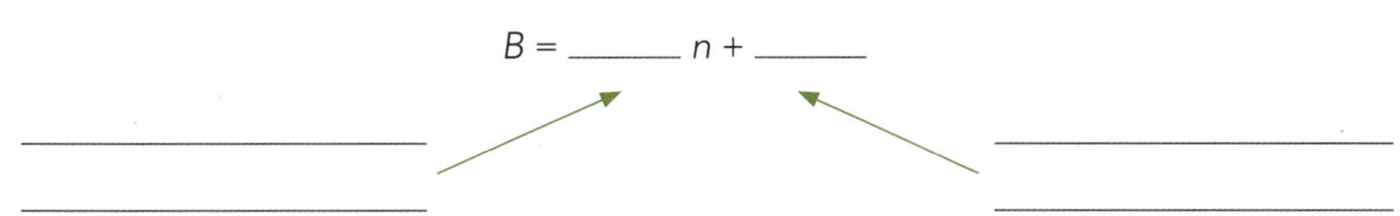

ISBN: 9780170370431

2 She makes another pattern with black and white buttons.

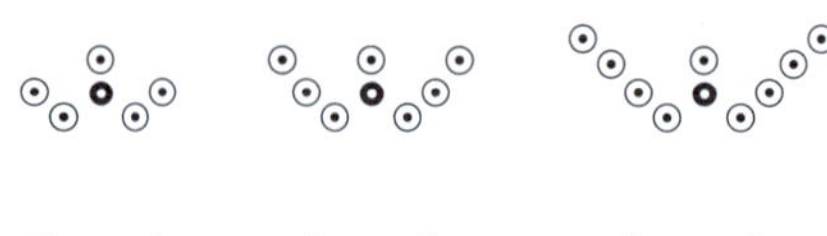

a Complete the table, and use it to find the equation.

Pattern # (n)	# of buttons (B)
1	6
2	8
3	10
4	
5	
6	

Equation: B = ______ n + ______

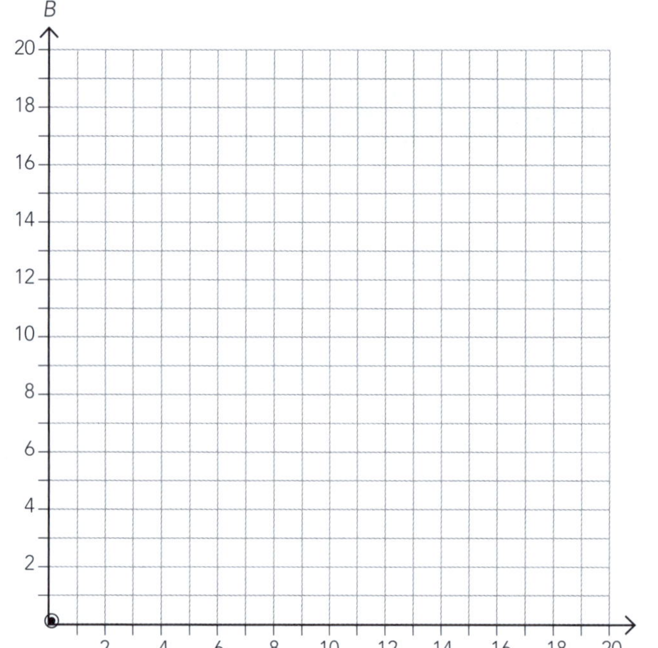

b Plot the points on the graph.

c How many buttons are needed for the 30th pattern?

30th pattern $\Rightarrow n = 30 \therefore B$ = ______ x 30 + ______ = ______.

So the 30th pattern needs ______ buttons.

d Which pattern would have 164 buttons?

164 buttons $\Rightarrow$ 164 = ______ n + ______

____

So the ______ pattern would need 164 buttons.

e How does the equation relate to the pattern?

B = ______ n + ______

____________________ ____________________

____________________ ____________________

 ISBN: 9780170370431

3 Nick makes the following pattern with matches. How many matches are needed for the 25th pattern?

Pattern 1

Pattern 2

Pattern 3

a Complete the table, and use it to find the equation.

Pattern # (n)	# of matchsticks (M)
1	5
2	9
3	13
4	
5	
6	

Equation: $M =$ ______ $n +$ ______

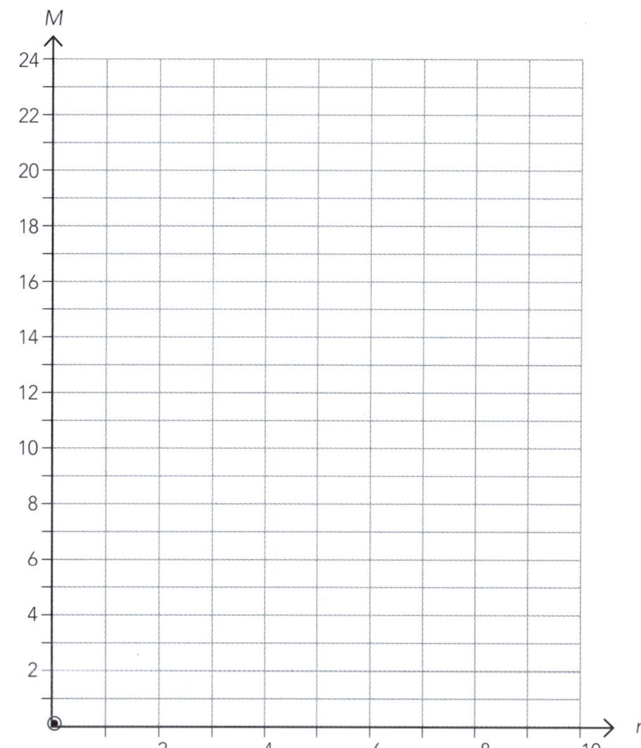

b Plot the points on the graph.

c How many matchsticks are needed for the 30th pattern?

30th pattern $\Rightarrow n = 30 \therefore M =$ ______ x 30 + ______ = ______.

So the 30th pattern needs ______ matchsticks.

d Which pattern would have 245 matchsticks?

245 matchsticks $\Rightarrow 245 =$ ______ $n +$ ______

____

So the ______ pattern would need 245 matchsticks.

e How does the equation relate to the pattern?

$M =$ ______ $n +$ ______

______________ ______________

______________ ______________

4 Tahu's dad will give him $40 if he passes NCEA Level 1, plus $10 for each standard in which he earns Merit or Excellence. The table below shows how much Tahu will get for his first few Merit or Excellence standards, assuming he passes Level 1.

a Complete the table, and use it to find the equation.

# of Merit or Excellence standards (s)	$ earned (D)
1	50
2	
3	
4	
5	
6	

Equation: D = ______ s + ______

b Plot the points on the graph.

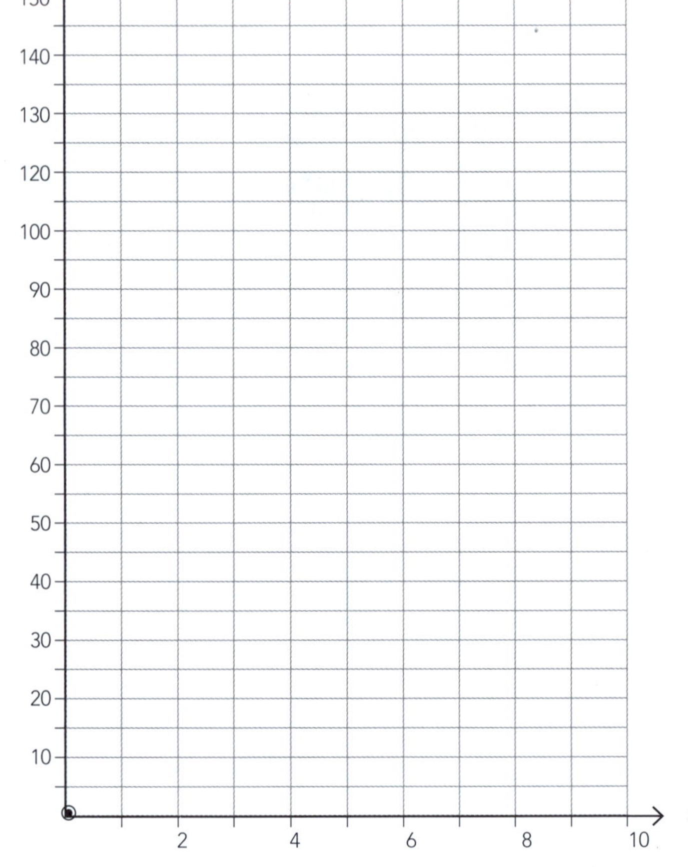

c How much does he earn if he passes 13 standards with Merit or Excellence?

13th pattern ⇒ $n = 13$ ∴ D = ______ x 13 + ______ = ______.

So if he passes 13 standards with Merit or Excellence, he earns $ ______.

d If he earns $210, how many standards did he pass with Merit or Excellence?

210 standards ⇒ 210 = ______ s + ______

____

So if he earned $210, he passed ______ standards with Merit or Excellence.

e How does the equation relate to the pattern?

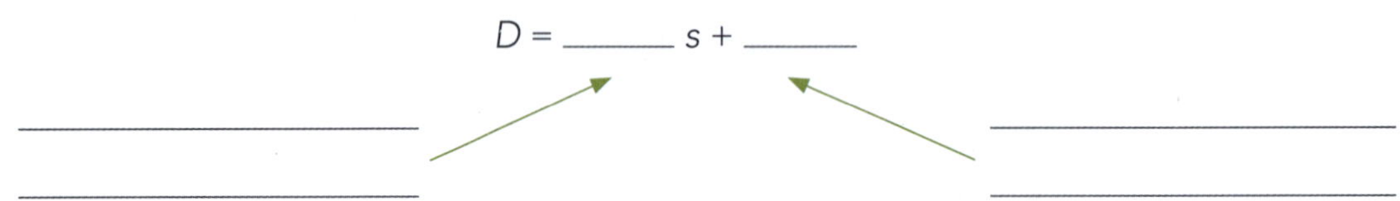

 ISBN: 9780170370431

5 Anna is planning a wedding breakfast at which guests will be seated at hexagonal tables arranged to form larger tables as shown.

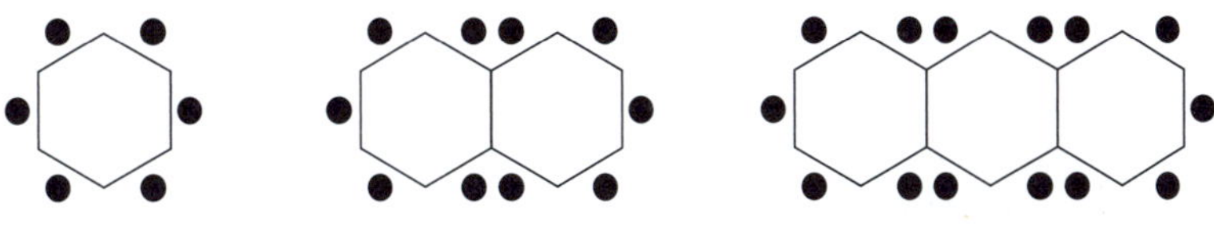

a Complete the table, and use it to find the equation.

# of tables (T)	# of guests (G)
1	
2	
3	
4	
5	
6	

Equation: G = ______ T + ______

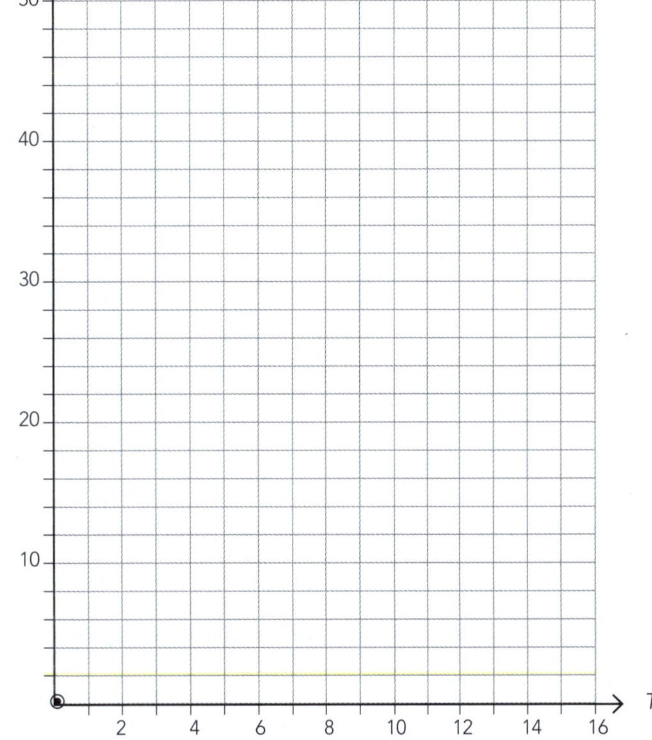

b Plot the points on the graph.

c How many guests could sit at 12 tables in a line?

____

____

Show where this value is on your graph. What do you notice about it?

d How many tables would be needed for 42 guests?

____

____

e How does the equation relate to the pattern?

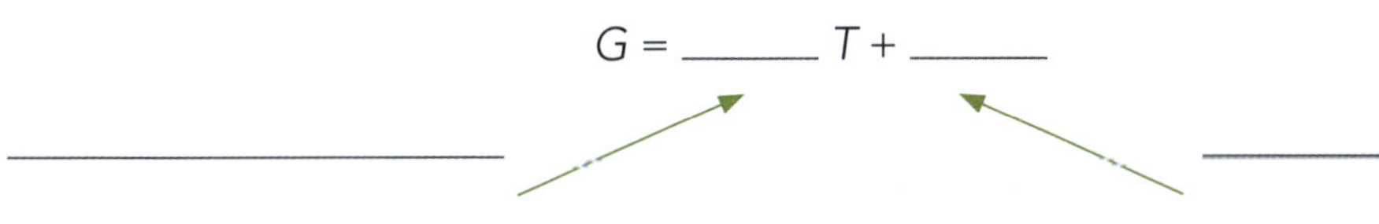

6 Matt has borrowed \$390 from his mum to buy a phone. He is paying her back \$15 at the end of each week.

a Complete the table, and use it to find the equation.

Week # (W)	\$ owed (D)
1	375
2	
3	
4	
5	
6	

Equation: D = ______ W + ______

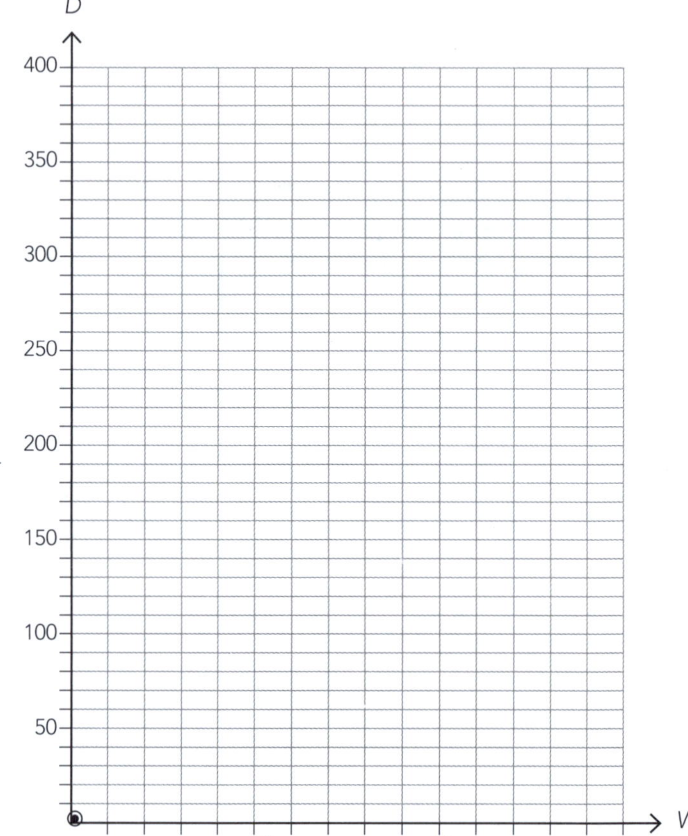

b Plot the points on the graph.

c How much does he owe at the end of the 12th week?

____

____

Show how you could find this value on your graph.

d At the end of which week has he repaid his mother completely?

____

____

Show how you could find this value on your graph.

e How does the equation relate to the pattern?

D = ______ W + ______

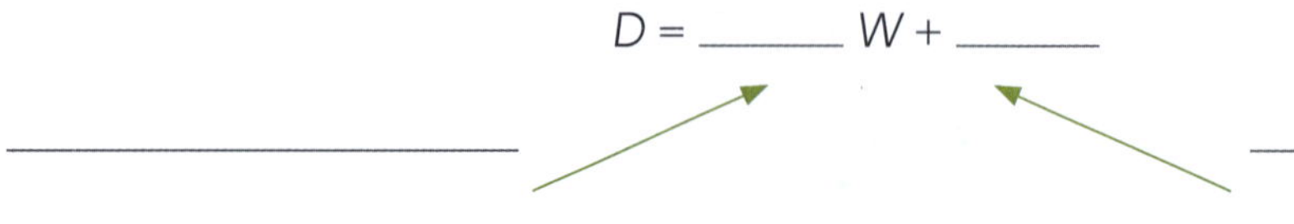

 ISBN: 9780170370431

The gradient of a line

The gradient is the **steepness**, or **slope**, of a line.

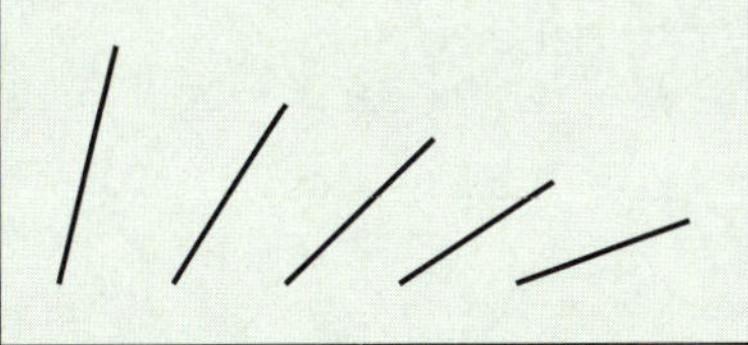

These lines all have **positive** gradients.

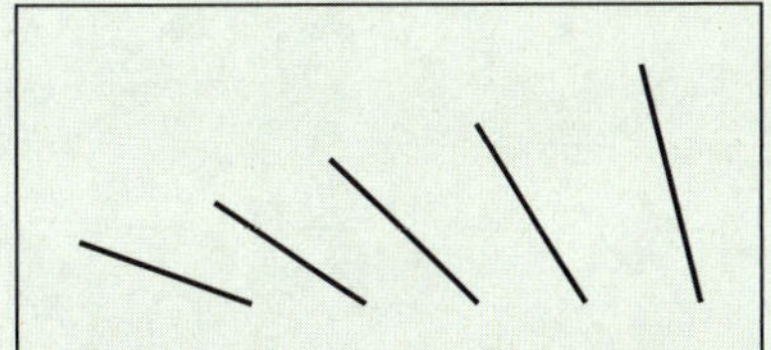

These lines all have **negative** gradients.

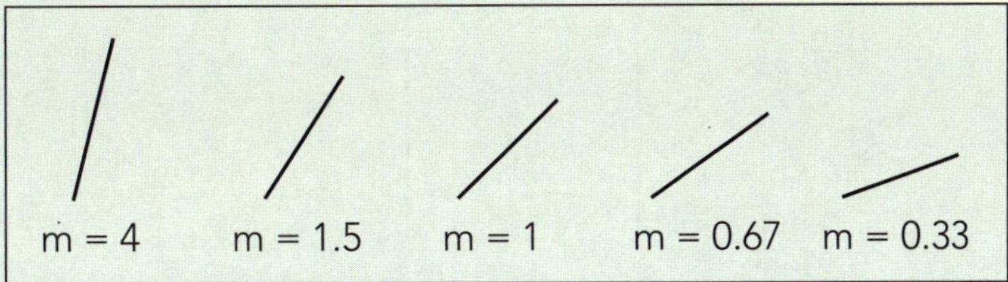

The **steeper** the line, the **bigger** the gradient.

The gradient is calculated using the formula $\mathbf{m = \frac{change\ in\ \mathit{y}}{change\ in\ \mathit{x}}}$ **or** $\mathbf{\frac{rise}{run}}$.

The easiest way to do this is to draw a right-angled triangle on the line.

$$m = \frac{rise}{run}$$

$$= \frac{5}{6}$$

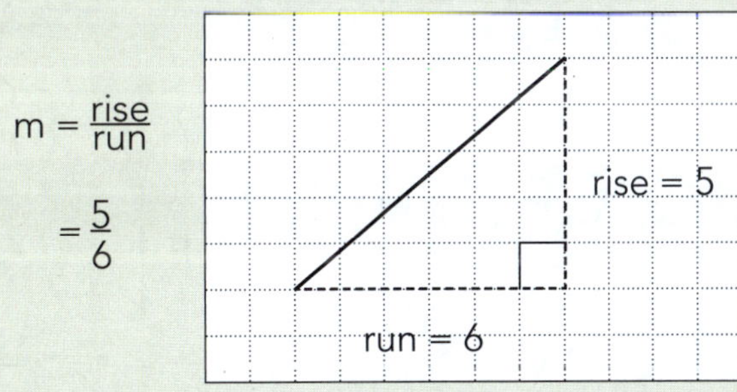

This time the gradient is *negative.*

$$m = -\frac{rise}{run}$$

$$= -\frac{3}{7}$$

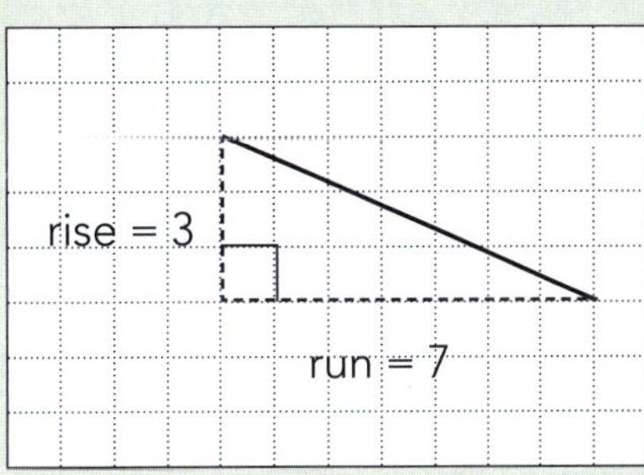

ISBN: 9780170370431

Horizontal lines

$$m = \frac{\text{rise}}{\text{run}}$$

$$= \frac{0}{7}$$

$$= 0$$

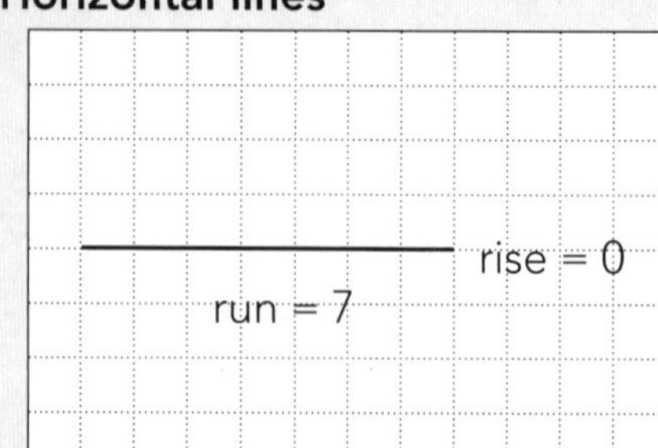

Vertical lines

$$m = \frac{\text{rise}}{\text{run}}$$

$$= \frac{5}{0}$$

$$= \text{undefined}$$

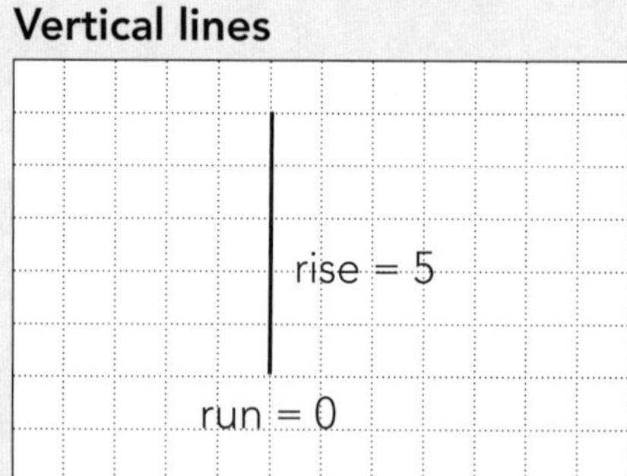

1 Calculate the gradients of these lines.

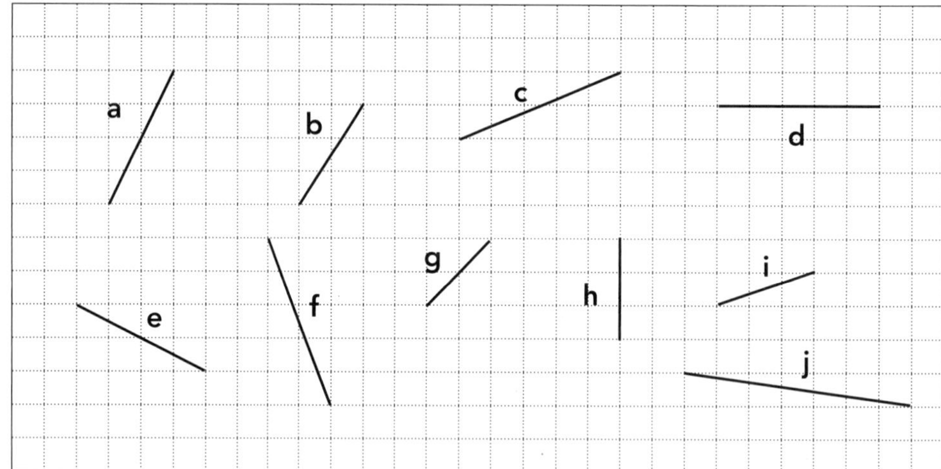

Gradient **a** = $\frac{\text{rise}}{\text{run}}$ = ____________________

Gradient **b** = ____________________

Gradient **c** = ____________________

Gradient **d** = ____________________

Gradient **e** = ____________________

Gradient **f** = ____________________

Gradient **g** = ____________________

Gradient **h** = ____________________

Gradient **i** = ____________________

Gradient **j** = ____________________

ISBN: 9780170370431

2 Draw line segments to show these gradients.

a $m = \frac{1}{3}$

b $m = 3$

c $m = \frac{3}{5}$

d $m = -2$

e $m = \frac{4}{7}$

f $m = -\frac{1}{4}$

g $m = -4$

h $m = -1\frac{1}{3}$

i $m = 0$

j $m = -1$

Gradients with different scales

- In practical situations, the scales on the axes are usually different.
- You need to be very careful when drawing or calculating these.

1 Drawing gradients

Example 1: m = 60, starting from the point (0, 100)

Step 1: Write the gradient as equivalent fractions:

$$m = \frac{\text{rise}}{\text{run}} = \frac{60}{1} = \frac{120}{2} = \mathbf{\frac{300}{5}} = \frac{600}{10} \text{ etc.}$$

Step 2: Select one that works with the given scales, and use it to draw another point. Join the points.

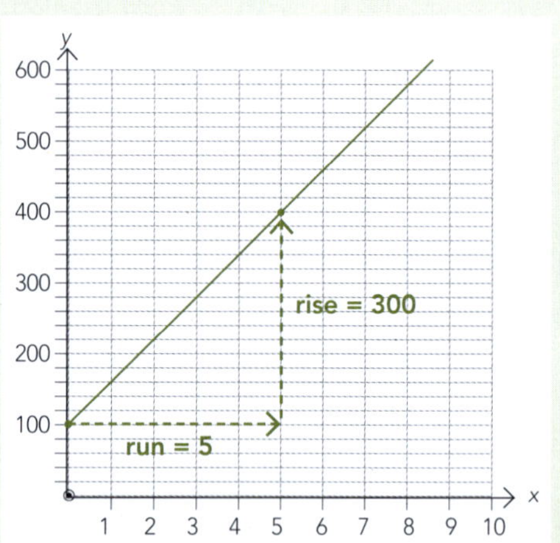

Example 2: $m = -\frac{2}{5}$, starting from the point (0, 28)

Step 1: Write the gradient as equivalent fractions:

$$m = \frac{\text{rise}}{\text{run}} = -\frac{2}{5} = -\frac{4}{10} = \mathbf{-\frac{20}{50}} \text{ etc.}$$

Step 2: Select one that works with the given scales, and use it to draw another point. Join the points.

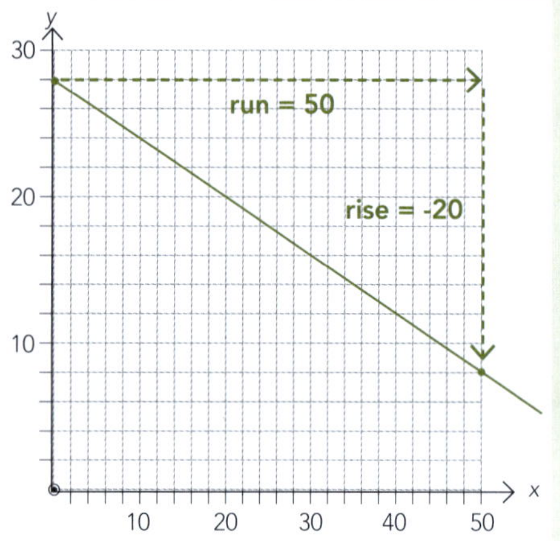

Draw the following gradients.

1 m = 40, starting from the point (0, 200)

$m = \frac{40}{1} =$

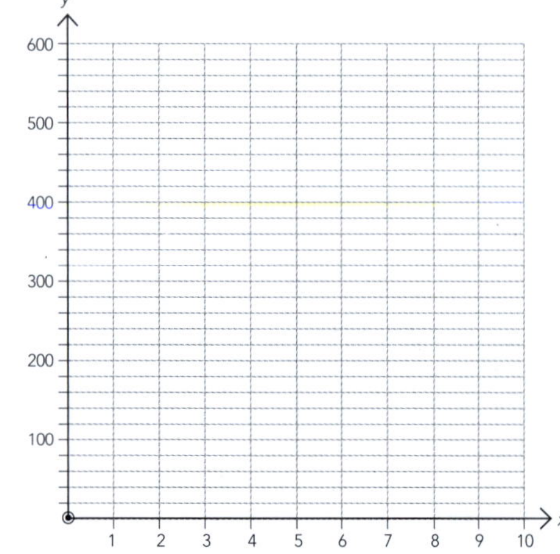

2 $m = -\frac{3}{5}$, starting from the point (0, 30)

$m = -\frac{3}{5} =$

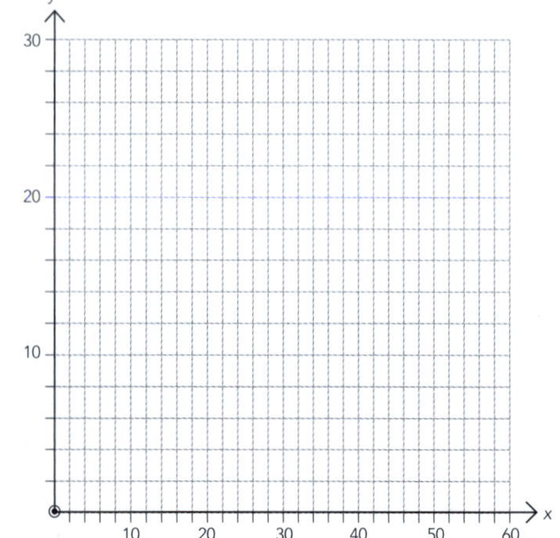

ISBN: 9780170370431

3 m = 3, starting from the point (0, 10)

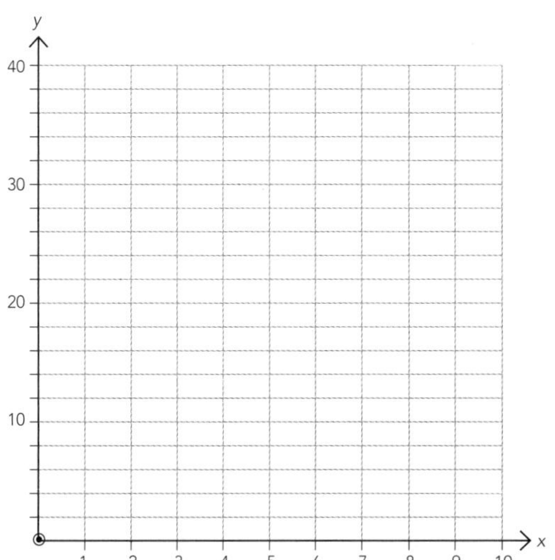

4 $m = -\frac{1}{2}$, starting from the point (0, 20)

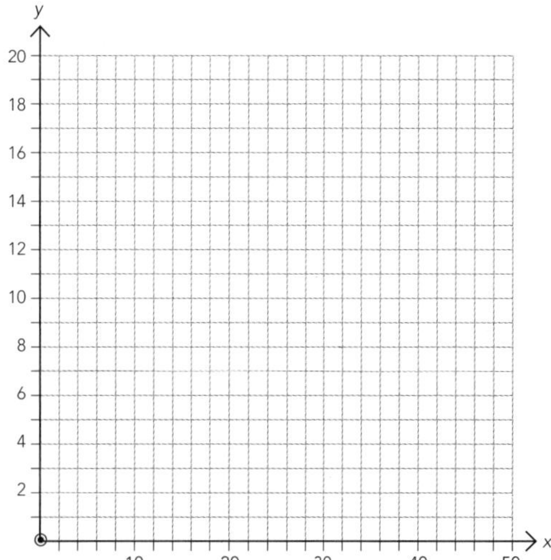

5 m = -6, starting from the point (0, 90)

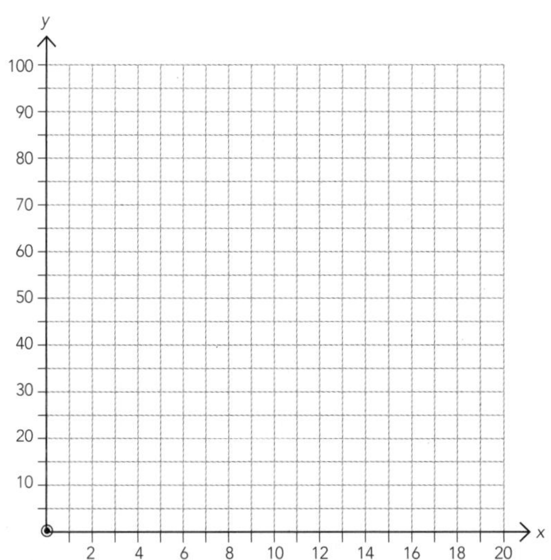

6 $m = \frac{1}{3}$, starting from the point (0, 4)

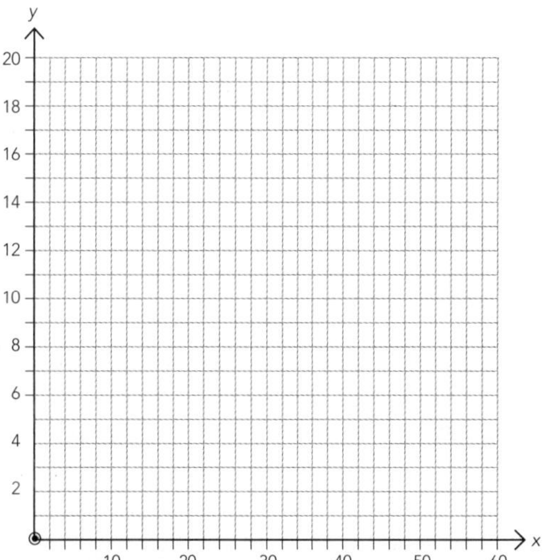

7 m = -2, starting from the point (0, 26)

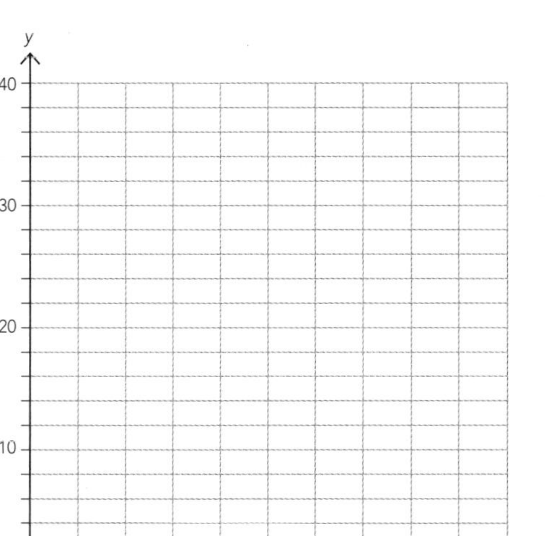

8 m = 12, starting from the point (0, 8)

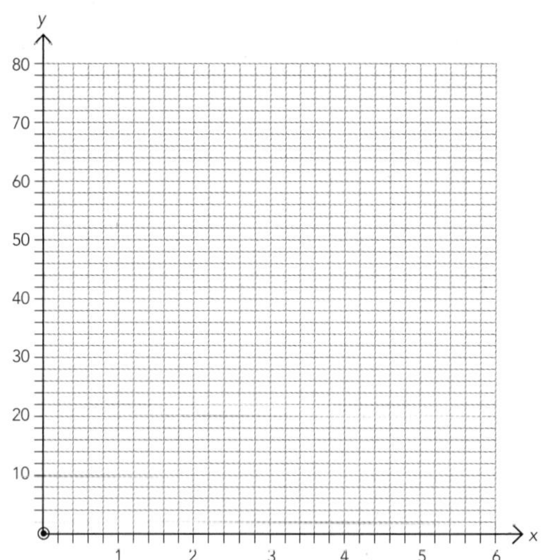

ISBN: 9780170370431

2 Calculating gradients

Example 1:

Step 1: Mark two points that the line clearly passes through and draw a right-angled triangle.

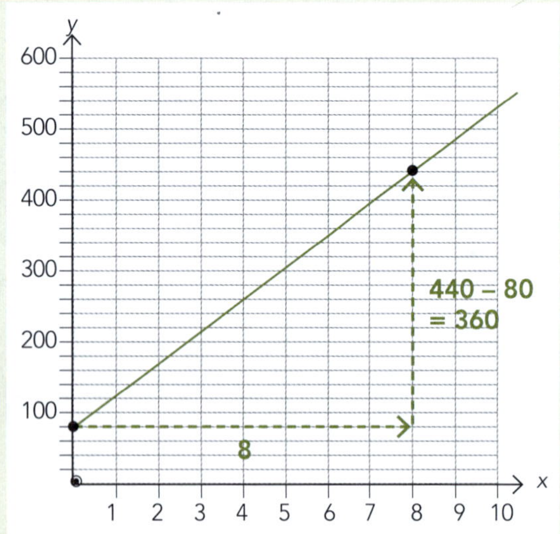

Step 2: Substitute:

$$m = \frac{\text{rise}}{\text{run}} = \frac{360}{8} = 45$$

Example 2:

Step 1: Mark two points that the line clearly passes through and draw a right-angled triangle.

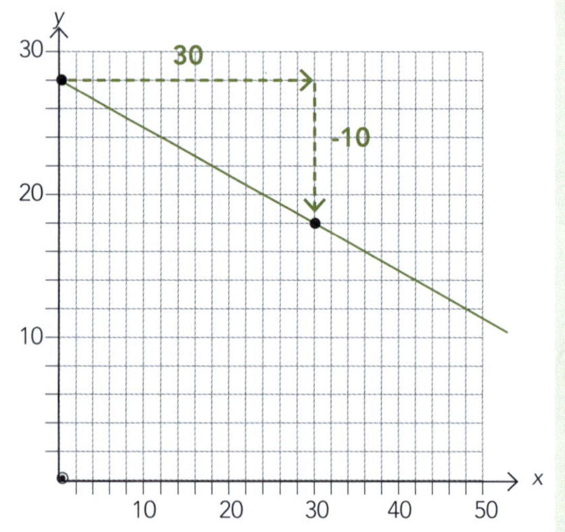

Step 2: Substitute:

$$m = \frac{\text{rise}}{\text{run}} = -\frac{10}{30} = -\frac{1}{3}$$

Calculate the following gradients.

1

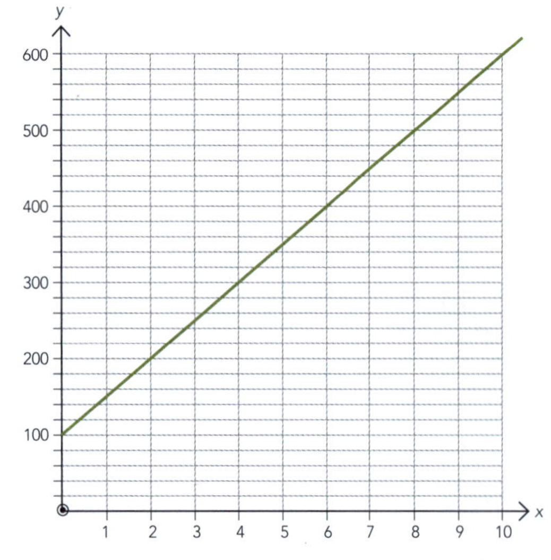

$m = \frac{\text{rise}}{\text{run}} = \frac{}{} =$

2

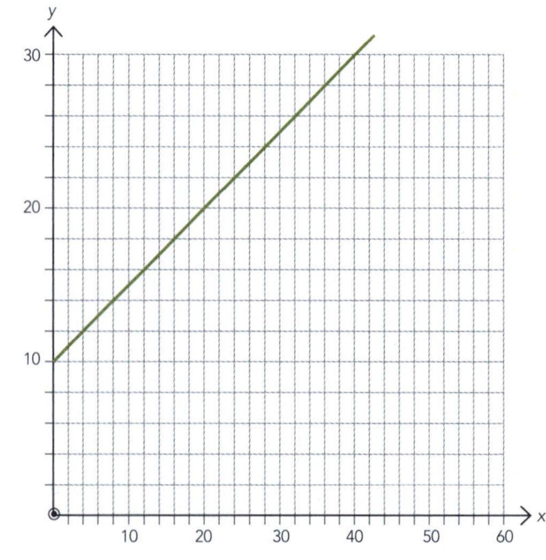

$m = \frac{\text{rise}}{\text{run}} = \frac{}{} =$

ISBN: 9780170370431

3

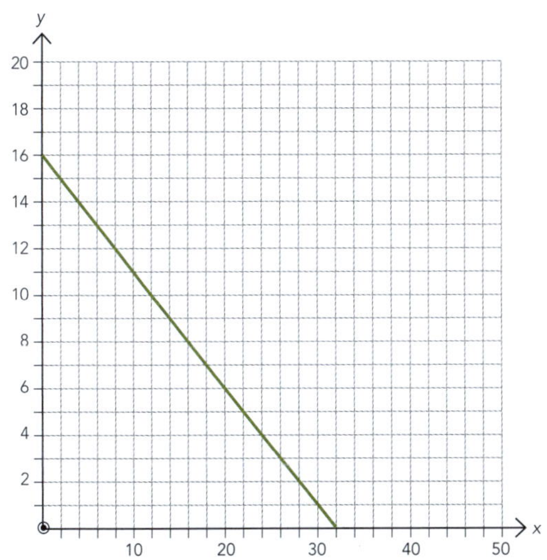

4

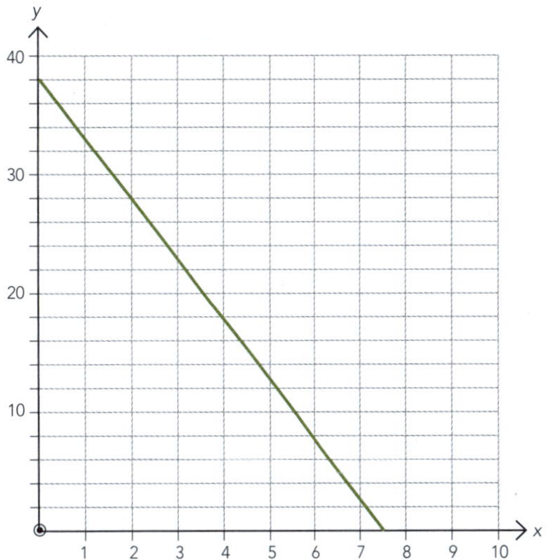

5

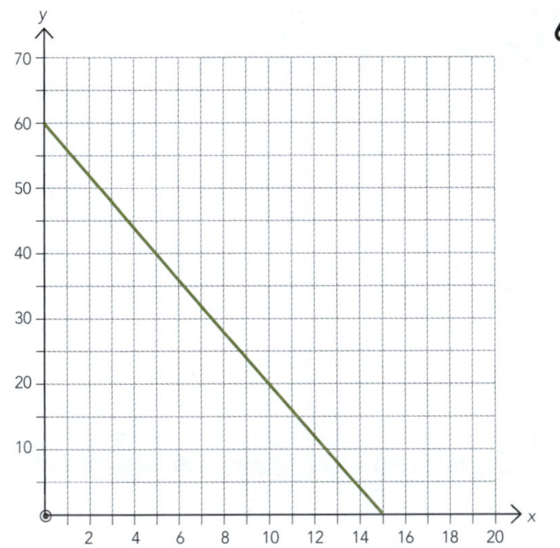

6

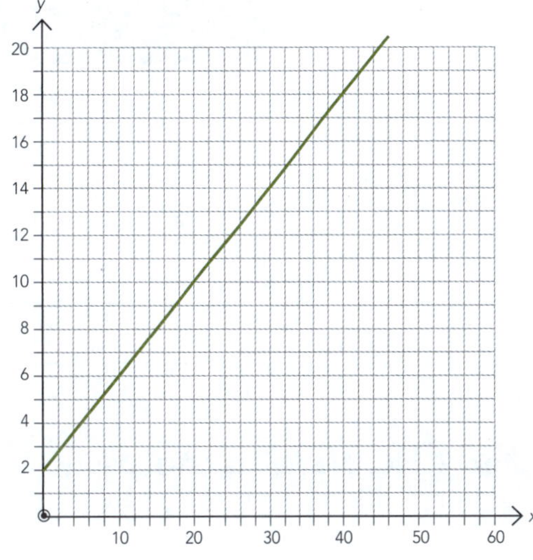

7

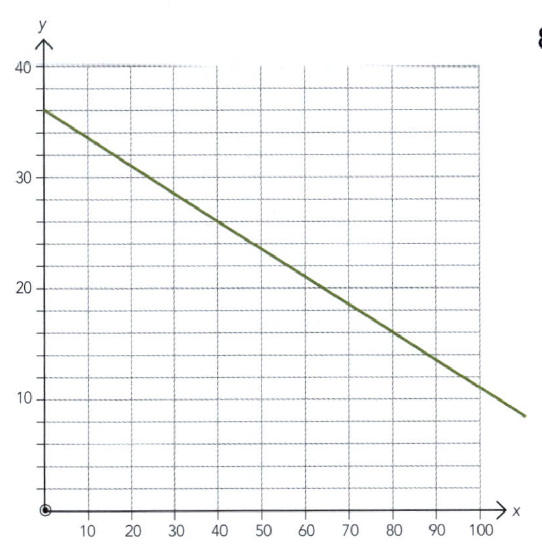

8

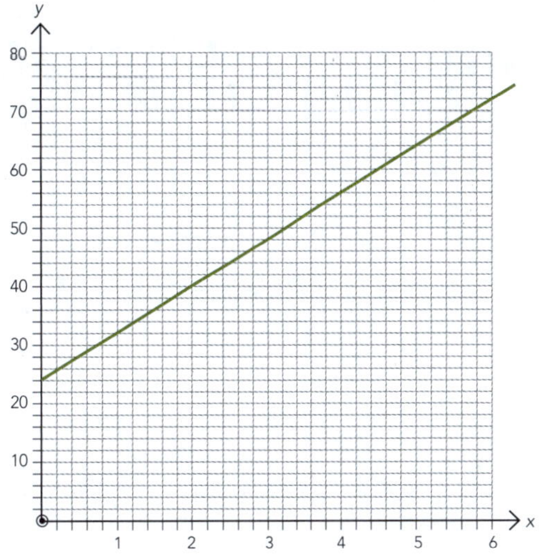

ISBN: 9780170370431

Drawing straight lines — continuous data

- **Continuous** data is unrounded **measured** data.
- It is shown as a **line** on a graph.

1 Plotting points using the equation

There are several ways of plotting graphs, but this method will work with *any* type of graph — lines and curves.

Example:
Plot the line given by the equation $y = 2x - 3$.

Step 1: Create a table for values of x and *y*.

x	2x – 3	y	Point
0	2(0) – 3	-3	(0, -3)
1	2(1) – 3	-1	(1, -1)
2	2(2) – 3	1	(2, 1)
3	2(3) – 3	3	(3, 3)
4	2(4) – 3	5	(4, 5)

Calculate at least three points.

Step 2: Plot the points on a graph.

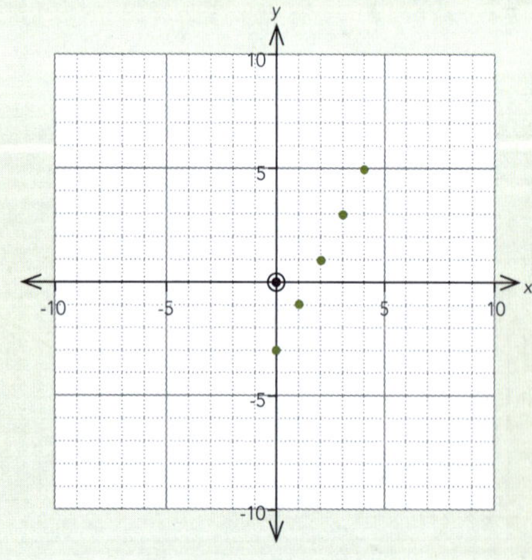

Step 3: Join the points with a **ruled** line.

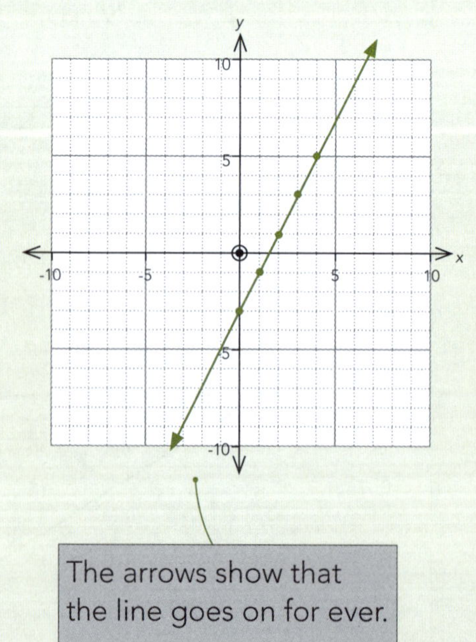

 ISBN: 9780170370431

Complete the tables and draw the graph for each of the following.

1 $y = x + 3$

x	x + 3	y	Point
0	(0) + 3	3	(0, 3)
1			
2			
3			
4			

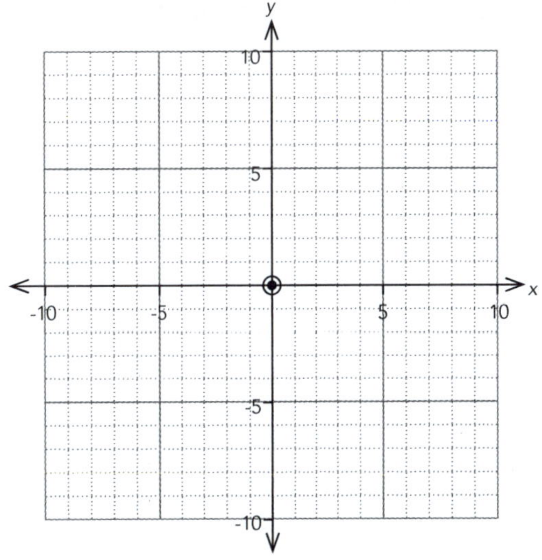

2 $y = 2x + 5$

x	2x + 5	y	Point
0			
1			
2			
3			
4			

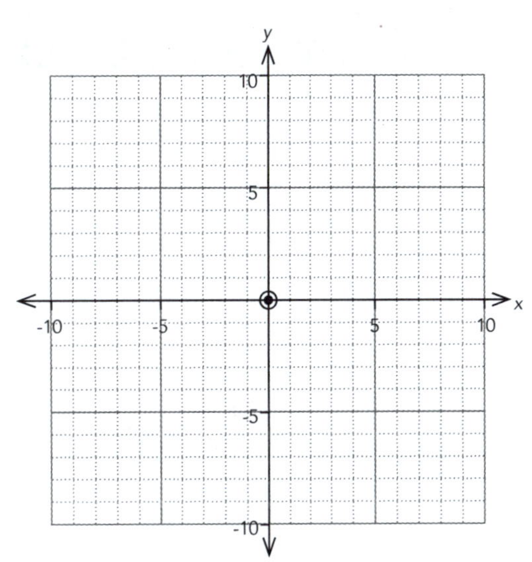

3 $y = 3x - 1$

x	3x − 1	y	Point
0			
1			
2			
3			
4			

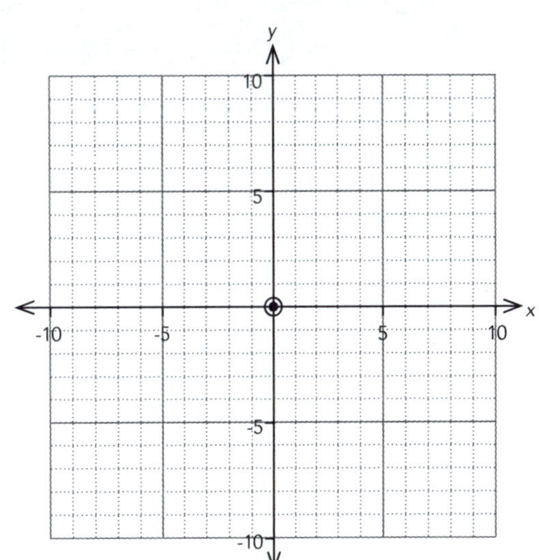

ISBN: 9780170370431

4 $y = -3x + 4$

x	-3x + 4	y	Point
0			
1			
2			
3			
4			

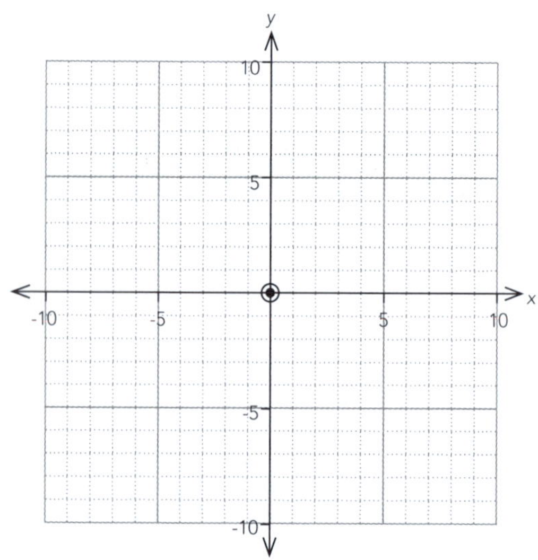

5 $y = 3$

x	3	y	Point
0			
1			
2			
3			
4			

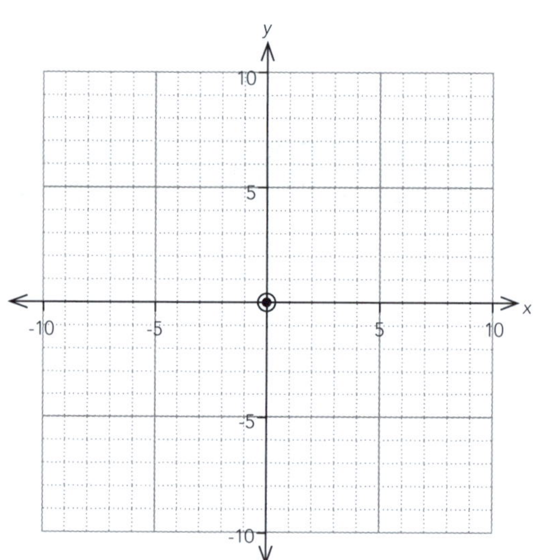

6 $y = -2x$

x	-2x	y	Point
0			
1			
2			
3			
4			

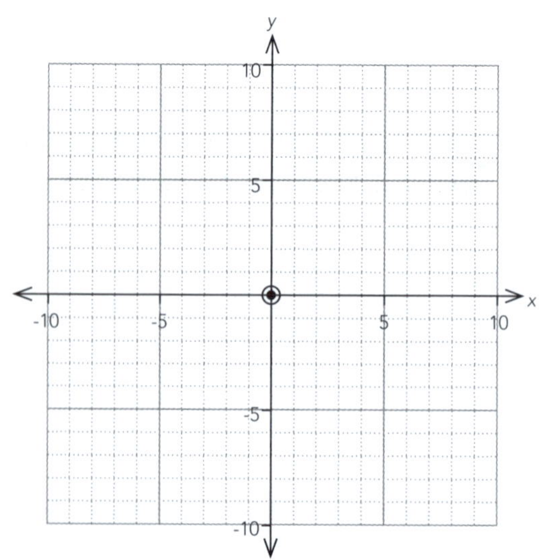

ISBN: 9780170370431

2 Using the *y*-intercept and the gradient

If you are given the equation of a graph, you can convert it into the form **y = mx + c** before drawing the graph.

m = gradient = $\frac{\text{rise}}{\text{run}}$ — **y = mx + c** — c = *y*-intercept

Examples:

1 Draw the graph of $y = -\frac{1}{2}x - 3$.

Step 1: Plot the *y*-intercept; in this case the *y*-intercept = **-3**.

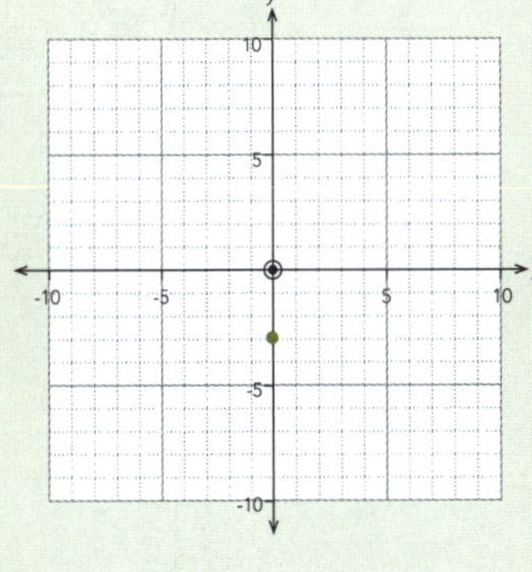

Step 2: From the intercept, plot at least three points along the gradient, in this case $\frac{-1}{2}$.

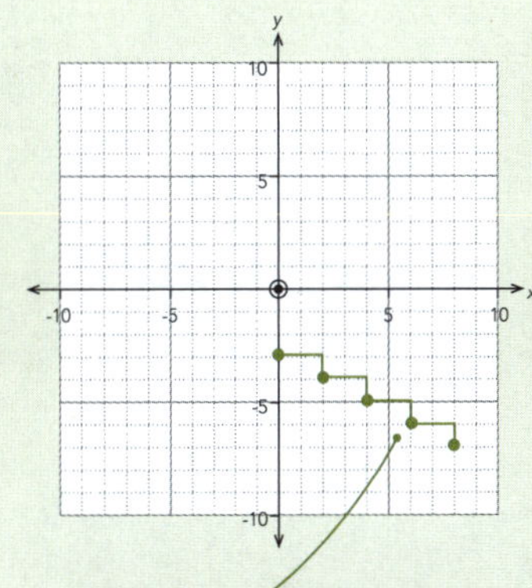

For each point, go across 2 and down 1.

Step 3: Join the points.

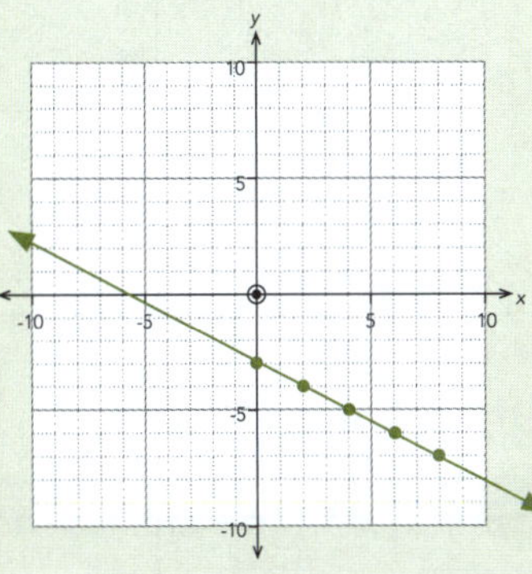

2 Draw the graph of $y = 60x + 200$.

Step 1: Plot the *y*-intercept; in this case the *y*-intercept = **200**.

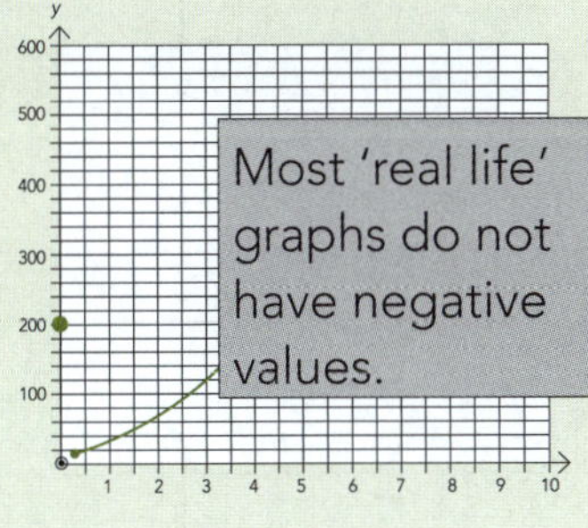

Most 'real life' graphs do not have negative values.

Step 2: From the intercept, plot at least three points along the gradient, in this case $\frac{60}{1}$.

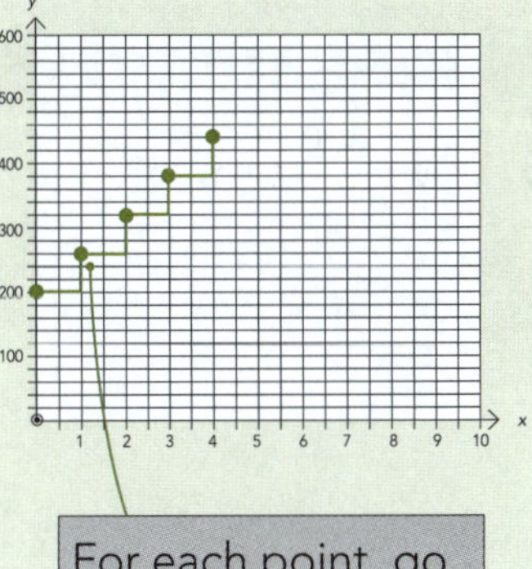

For each point, go across 1 and up 60.

Step 3: Join the points.

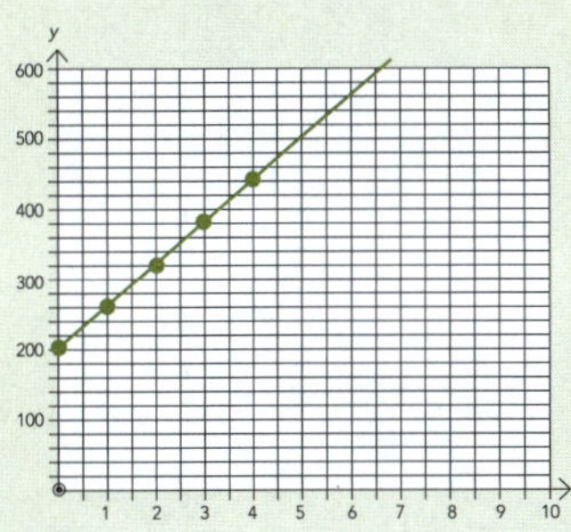

Draw the following lines.

1 $y = x + 4$

2 $y = 2x - 5$

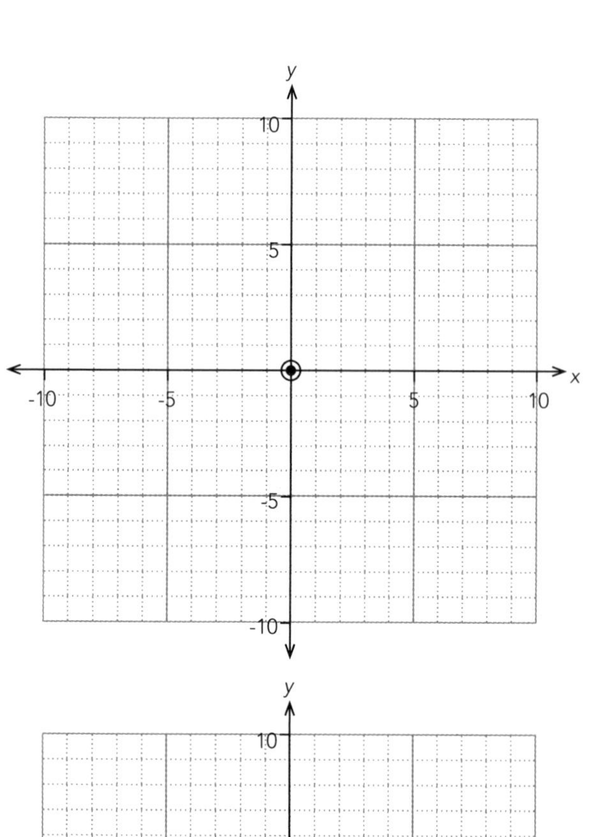

3 $y = -3x + 2$

4 $y = \frac{1}{3}x - 4$

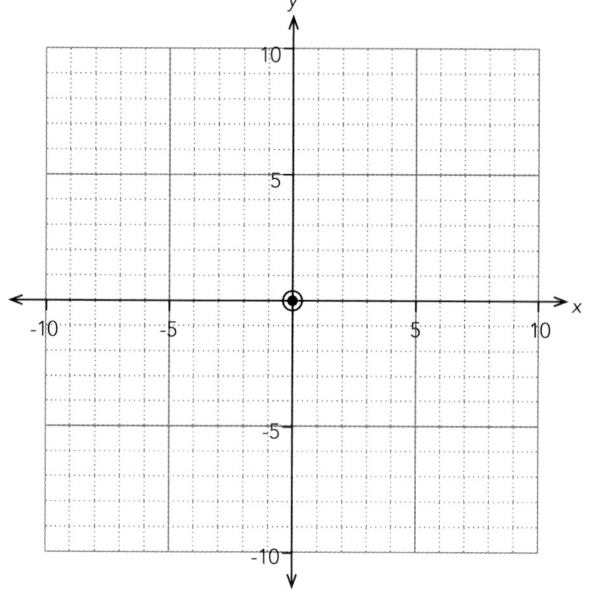

5 $y = -\frac{1}{4}x - 5$

6 $y = -3x$

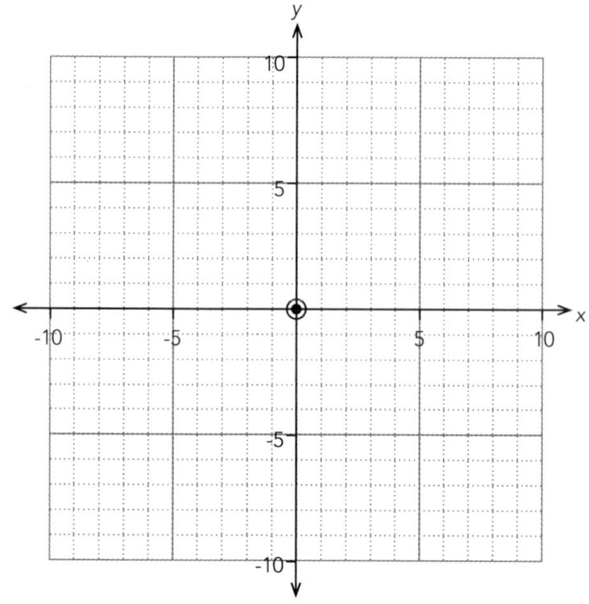

 ISBN: 9780170370431

7 $y = 100x + 200$

8 $y = -50x + 500$

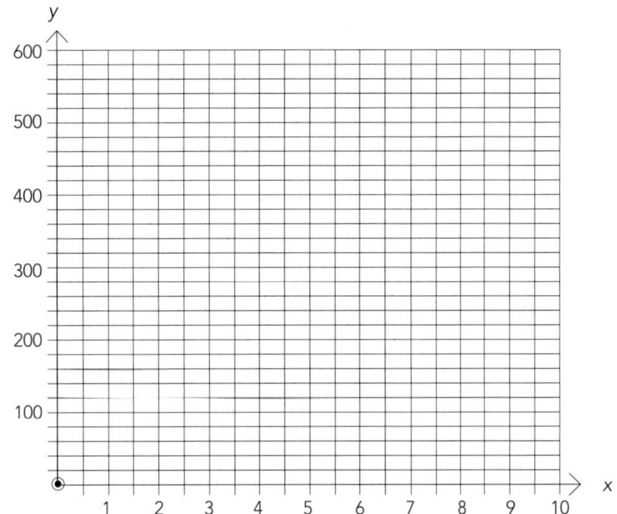

9 $y = 6x + 12$

10 $y = -4x + 36$

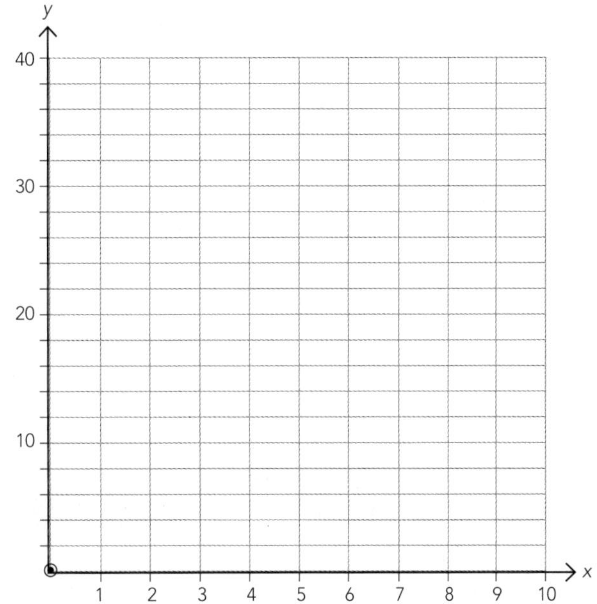

11 $y = -0.5x + 18$

12 $y = 0.2x + 5$

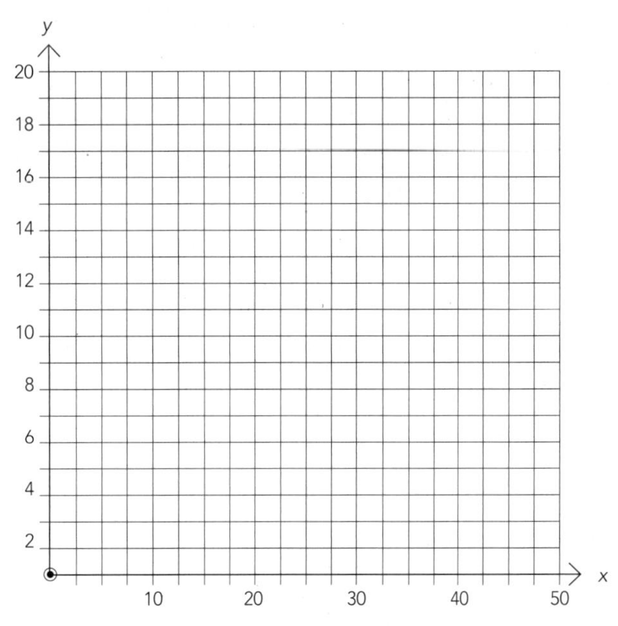

ISBN: 9780170370431

3 Using the *x*- and *y*-intercepts

- Often you will need to draw the graph of an equation that is not in $y = mx + c$ form.
- If it is in the form $ax + by = c$, it is usually easier to use the **intercept-intercept** method in order to draw the graph.

Examples:

1 Draw the graph of $6x - 3y = 12$.

Step 1: Make $x = 0$.
Then $6(0) - 3y = 12$
$\therefore\ y = \mathbf{-4}$
Plot the point (0, **-4**).

Step 2: Make $y = 0$.
Then $6x - 3(0) = 12$
$\therefore\ x = \mathbf{2}$
Plot the point (**2**, 0).

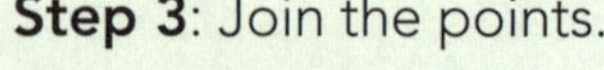
Step 3: Join the points.

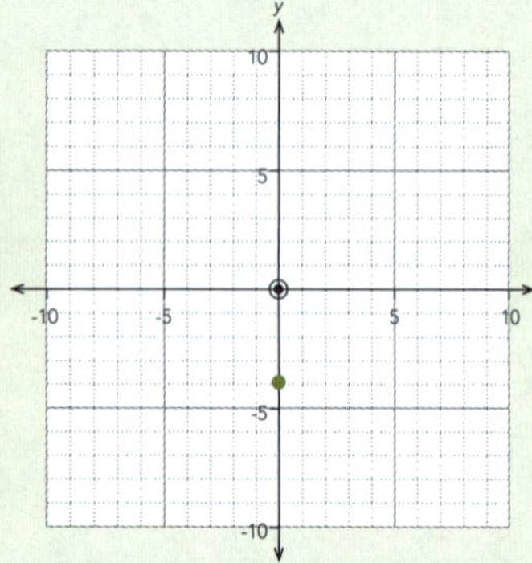

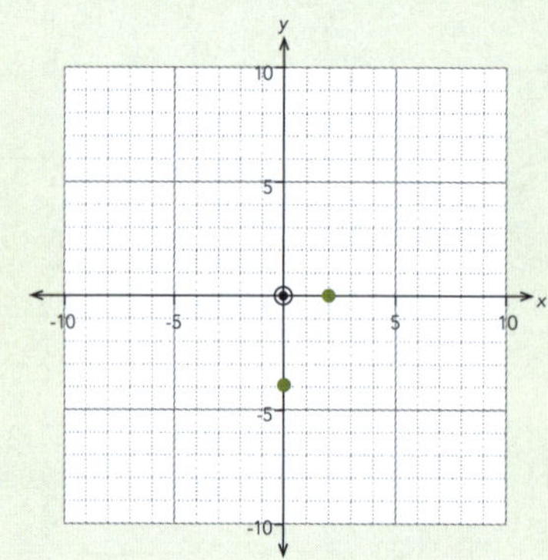

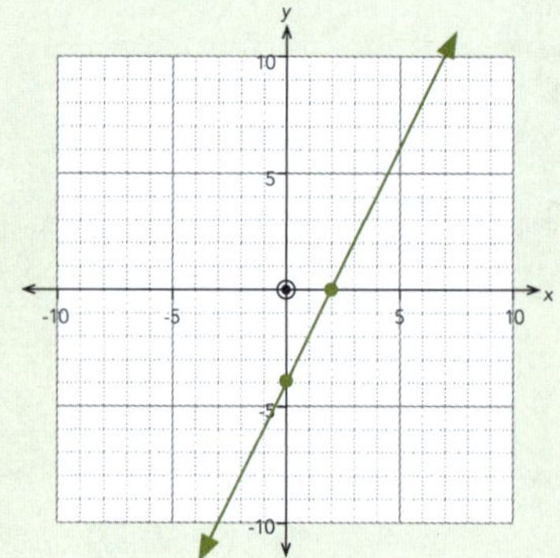

2 Draw the graph of $3x + 2y = 90$.

Step 1: Make $x = 0$.
Then $3(0) + 2y = 90$
$\therefore\ y = \mathbf{45}$
Plot the point (0, **45**).

Step 2: Make $y = 0$.
Then $3x + 2(0) = 90$
$\therefore\ x = \mathbf{30}$
Plot the point (**30**, 0).

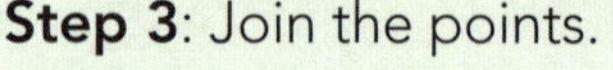
Step 3: Join the points.

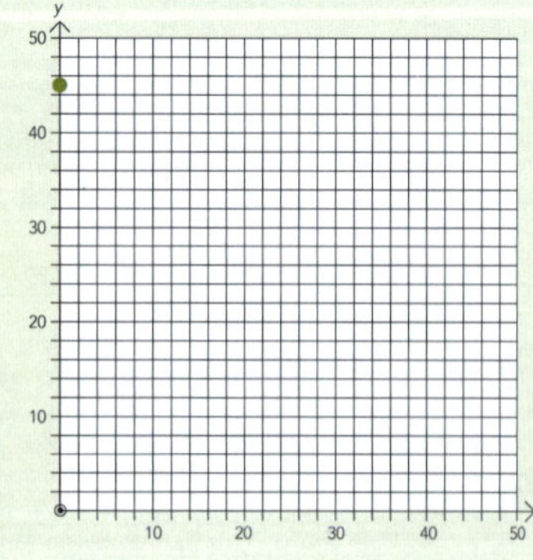

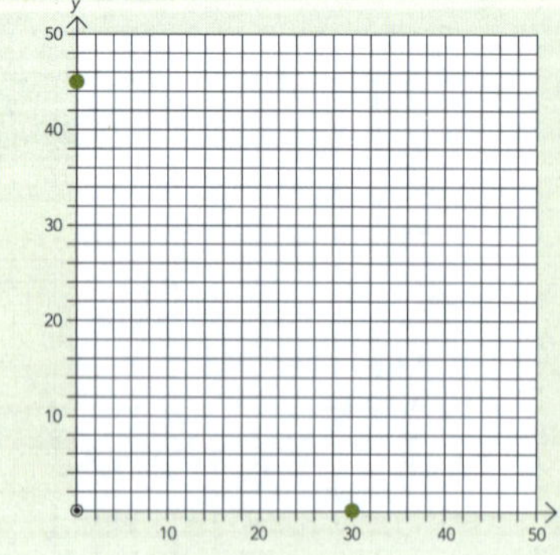

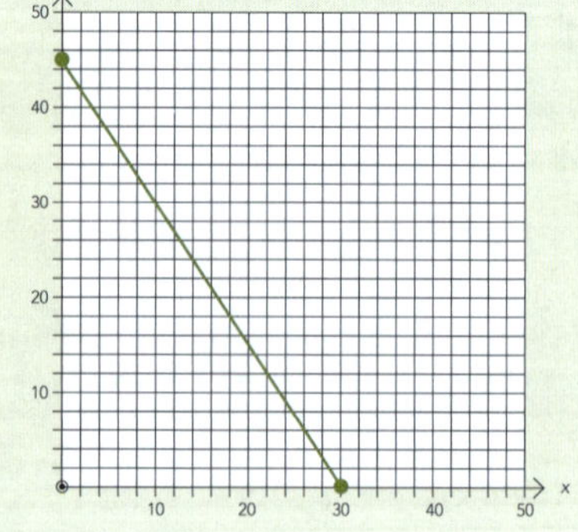

ISBN: 9780170370431

Draw the following lines.

1 $4x - 2y = 20$
$x = 0 \Rightarrow y =$ ____ $y = 0 \Rightarrow x =$ ____

2 $3x + 4y = 12$
$x = 0 \Rightarrow y =$ ____ $y = 0 \Rightarrow x =$ ____

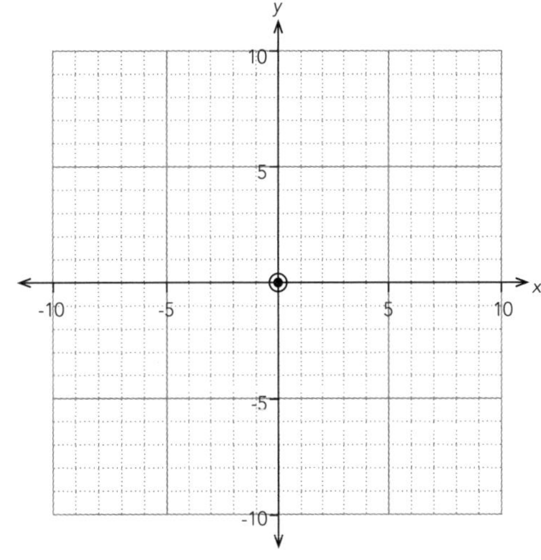

3 $3x + 2y = 18$

4 $3x - 6y = 24$

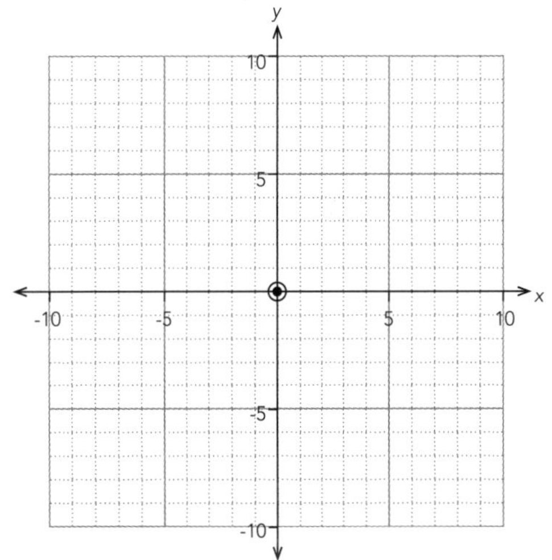

5 $3x + 2y = 90$

6 $4x + 5y = 200$

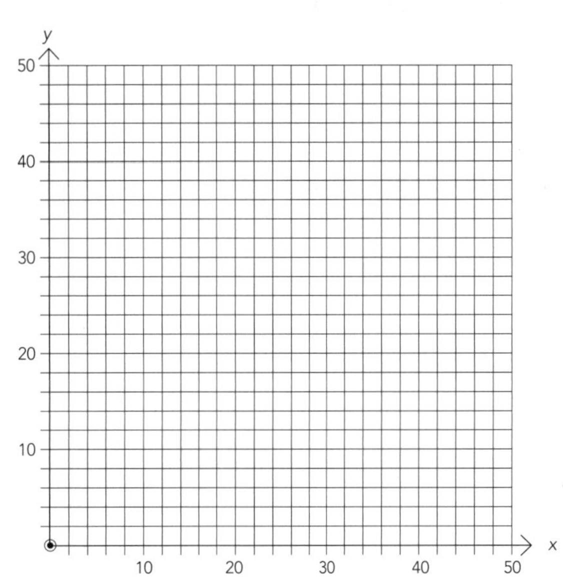

ISBN: 9780170370431

7 $4y + 9x = 180$

8 $y + 5x = 80$

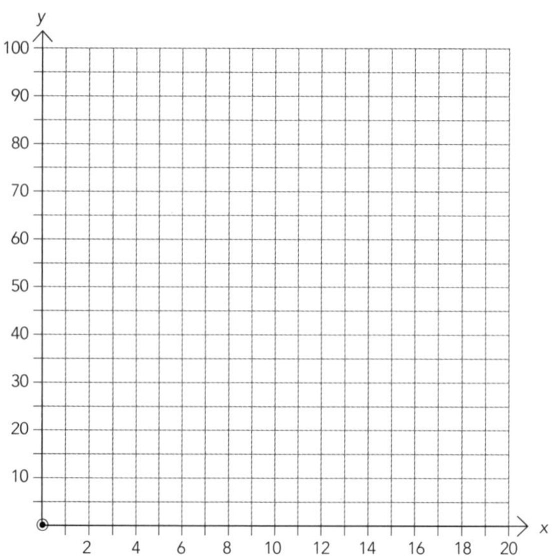

9 $30x + 14y = 210$

10 $32x + 5y = 160$

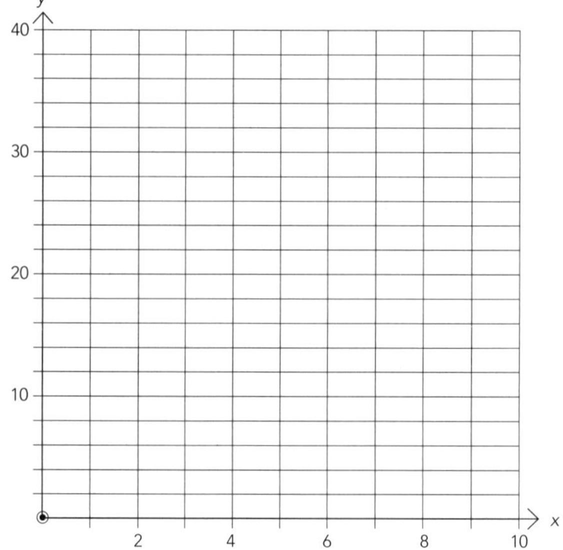

11 $8x + 25y = 400$

12 $6x + 18y = 360$

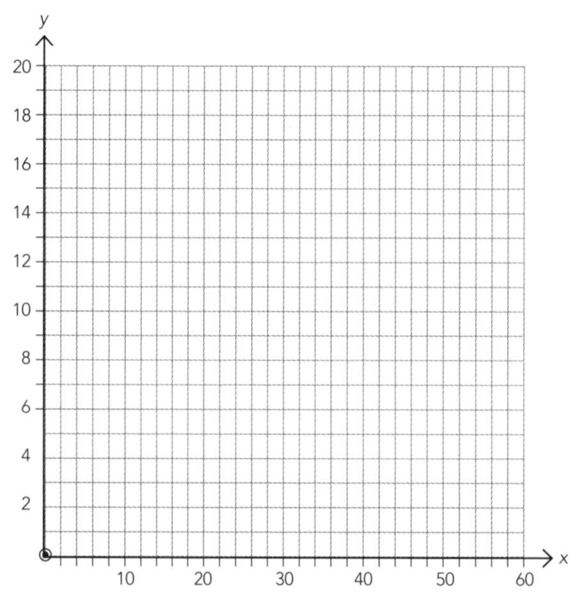

 ISBN: 9780170370431

4 Horizontal and vertical lines

Be very careful when drawing these — it's easy to get them the wrong way around.

Examples:

1 Draw the line $x = -3$.

Step 1: Plot at least three points where $x = -3$, e.g. (-3, 0), (-3, 5), (-3, -4).

Step 2: Join the points.

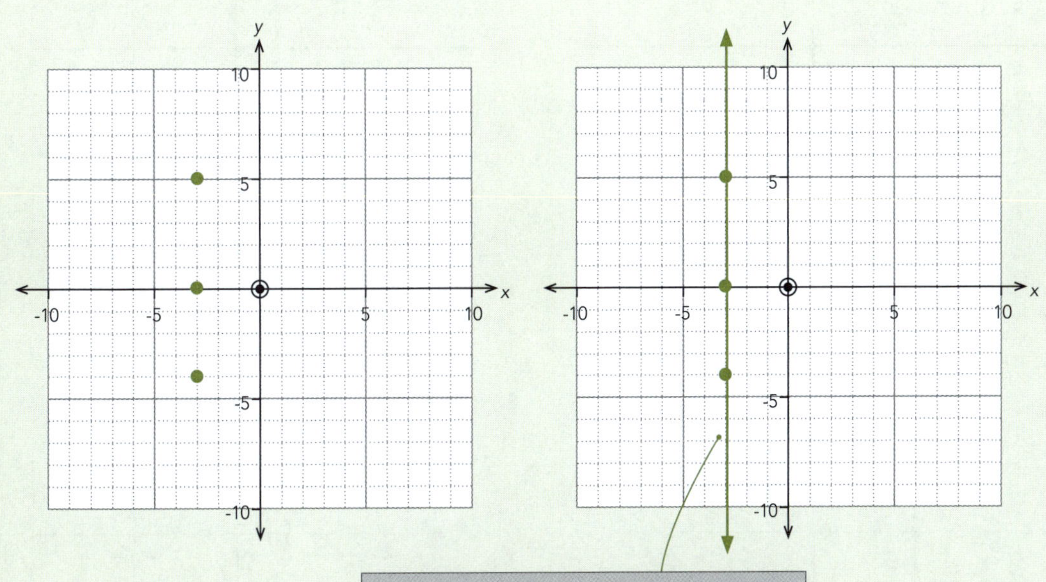

Notice that although this is **x** = -3, it is parallel with the **y**-axis.

2 Draw the line $y = 4$.

Step 1: Plot at least three points where $y = 4$, e.g. (0, 4), (2, 4), (-3, 4).

Step 2: Join the points.

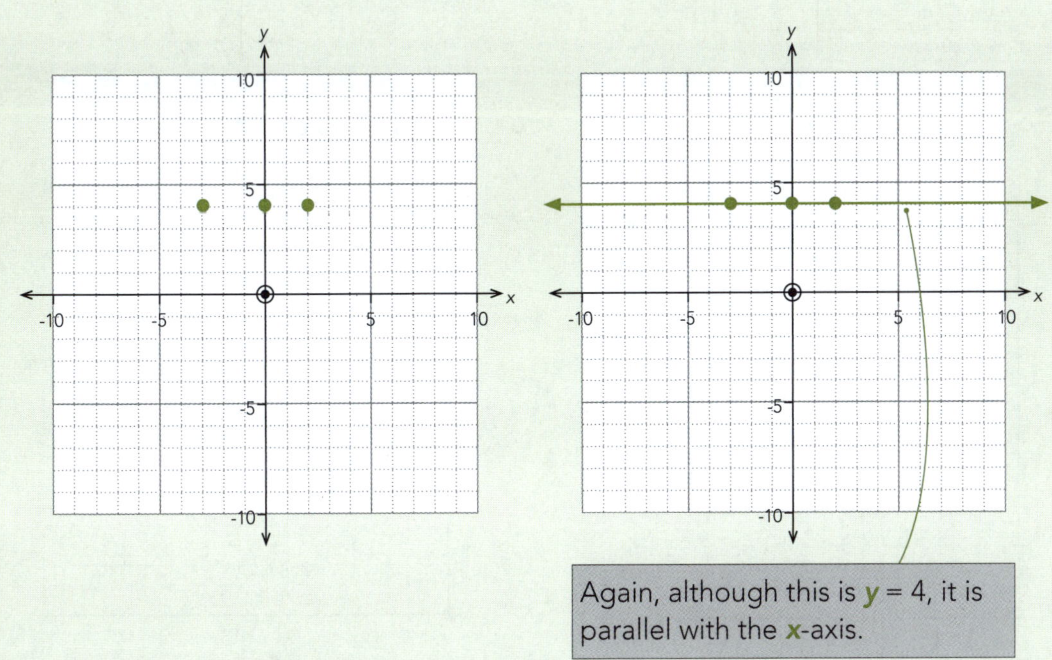

Again, although this is **y** = 4, it is parallel with the **x**-axis.

Draw the following lines.

1 $x = 3$ and $y = 5$

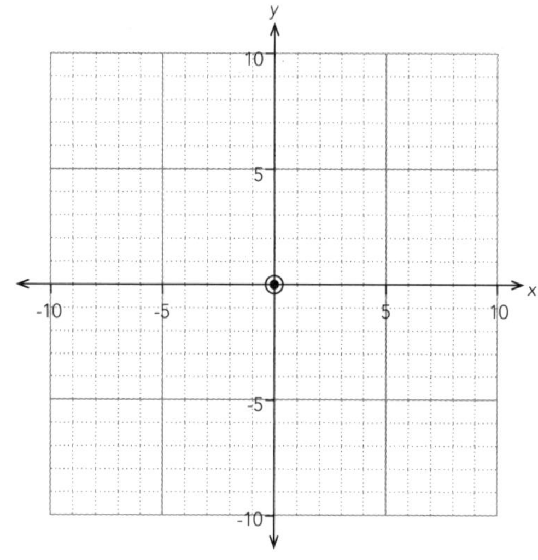

2 $y = -2$ and $x = 7$

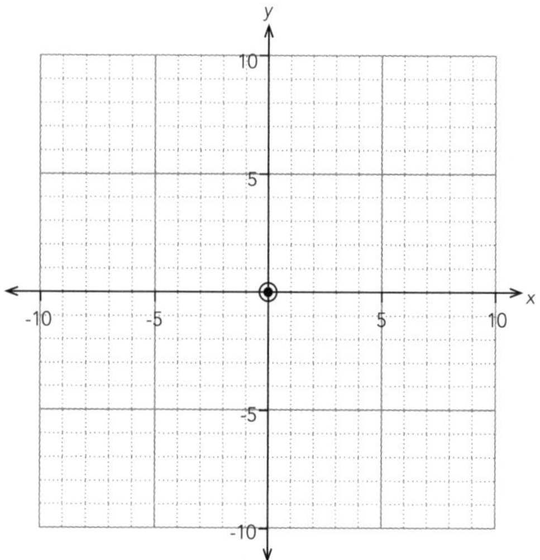

3 $x = -1$ and $y = 0$

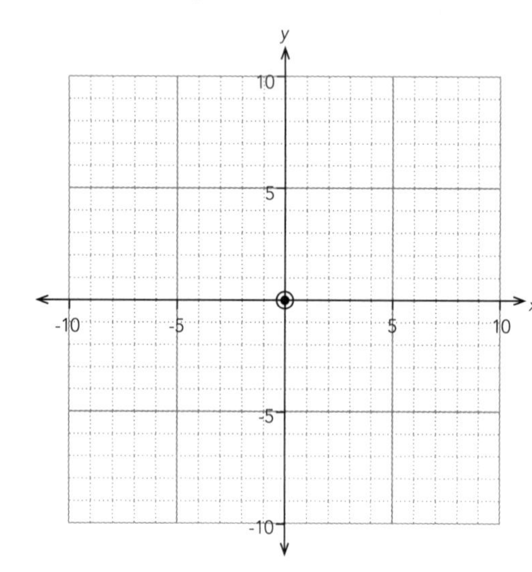

4 $y = 6$ and $x = 0$

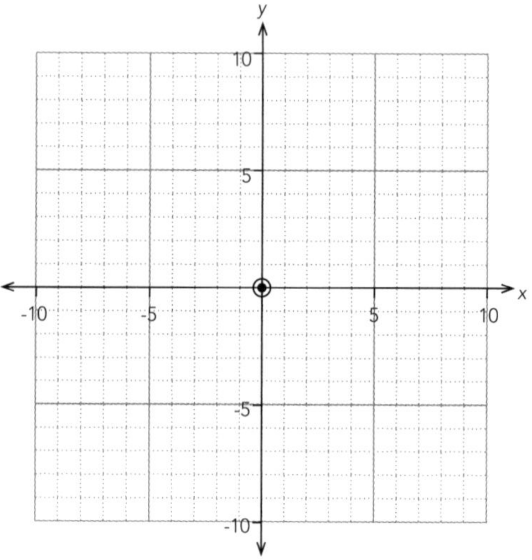

5 $b = 7.5$ and $y = 42$

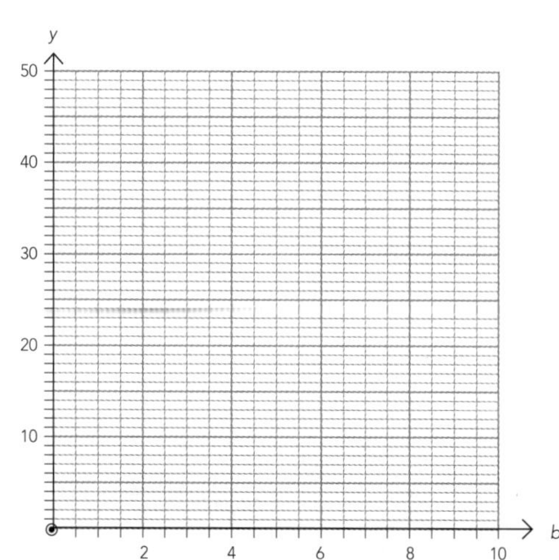

6 $T = 43$ and $S = 76$

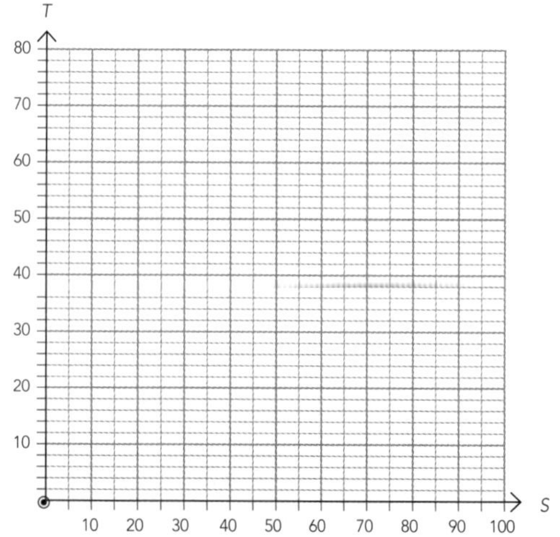

 ISBN: 9780170370431

Writing equations from graphs

1 Using the gradient and the *y*-intercept

Use **y = mx + c** as a template for your equation.

Example 1: Write the equation for this line.

Step 1: Find the *y*-intercept (c).

c = -4

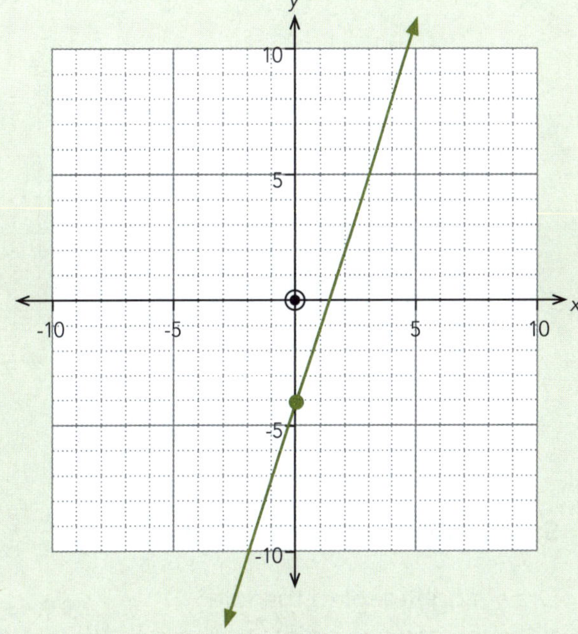

Step 2: Work out the gradient (m).

Draw a right-angled triangle between any two points through which the line passes.

$$m = \frac{\text{rise}}{\text{run}} = \frac{+12}{+4} = 3$$

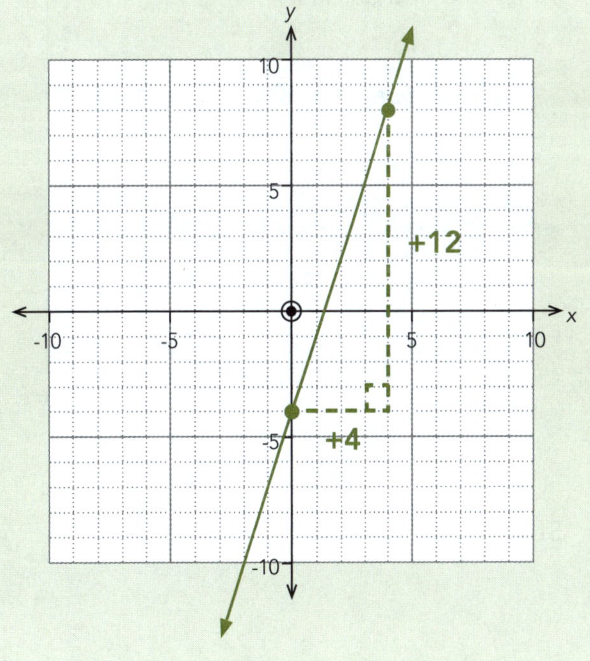

Step 3: Substitute into $y = mx + c$.

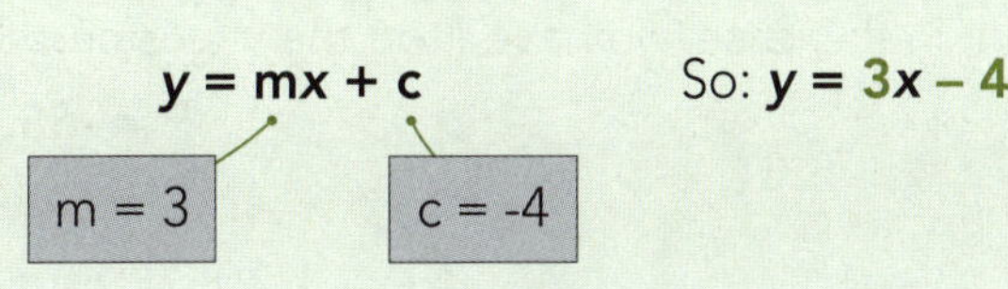

So: **y = 3x – 4**

ISBN: 9780170370431

Example 2: Write the equation for this line.

Step 1: Find the *y*-intercept (c).

c = 90

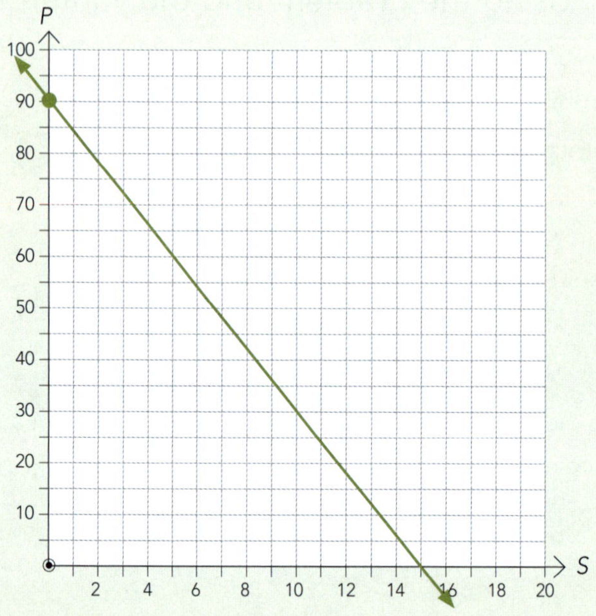

Step 2: Work out the gradient (m).

Draw a right-angled triangle between any two points through which the line passes.

$$m = \frac{\text{rise}}{\text{run}} = \frac{-60}{+10} = -6$$

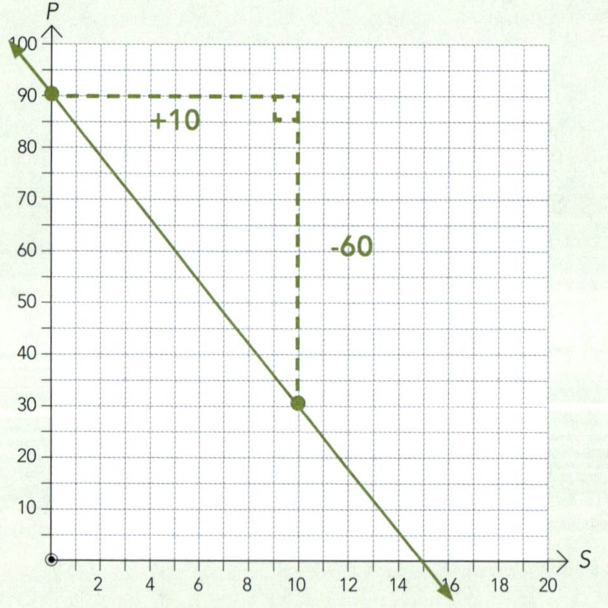

Step 3: Substitute into $y = mx + c$.

$y = mx + c$

m = -6

c = 90

So: **$y = -6x + 90$**

But, in this case, we have *P* on the y-axis and *S* on the *x*-axis. So the equation has to be:

$P = -6S + 90$

 ISBN: 9780170370431

Write equations for the following lines.

1

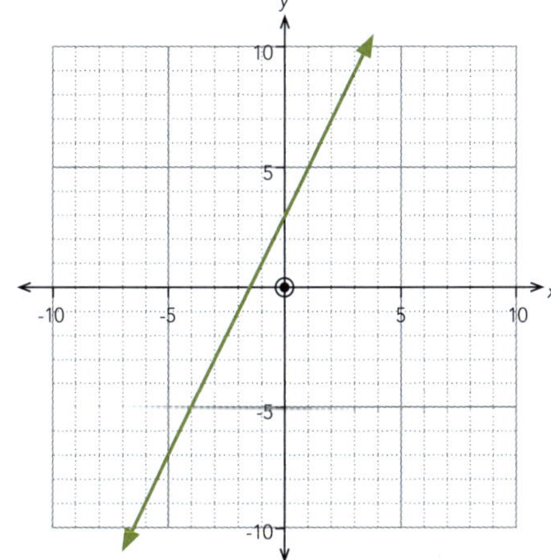

$y =$ ______ x ______

2

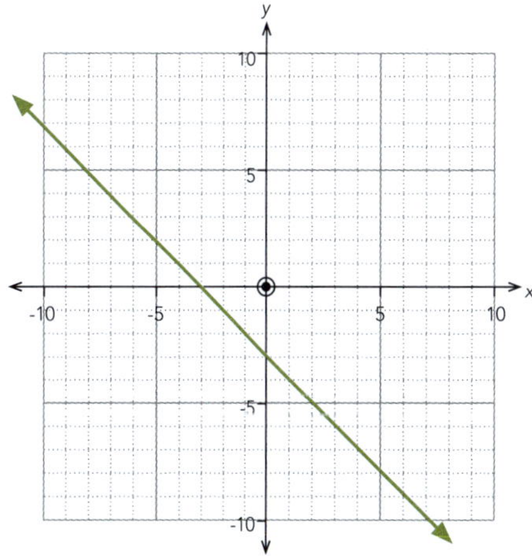

$y =$ ______ x ______

3

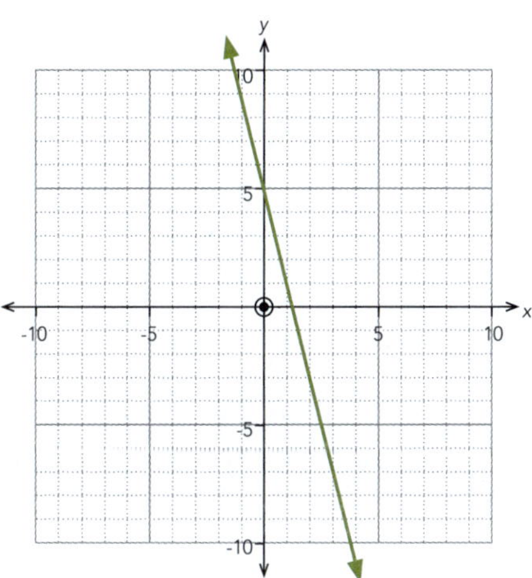

$y =$ ____________________

4

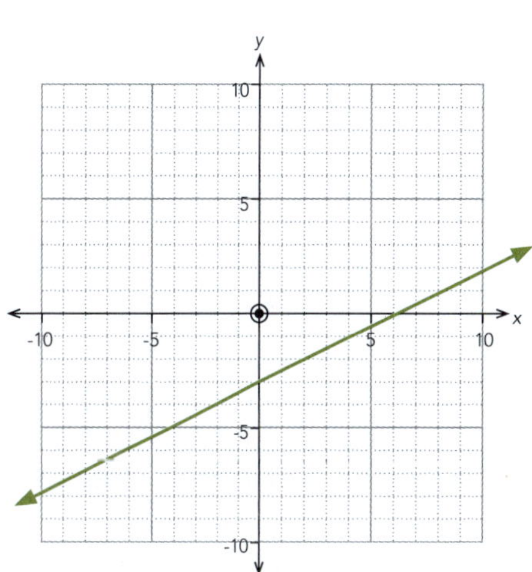

$y =$ ____________________

5

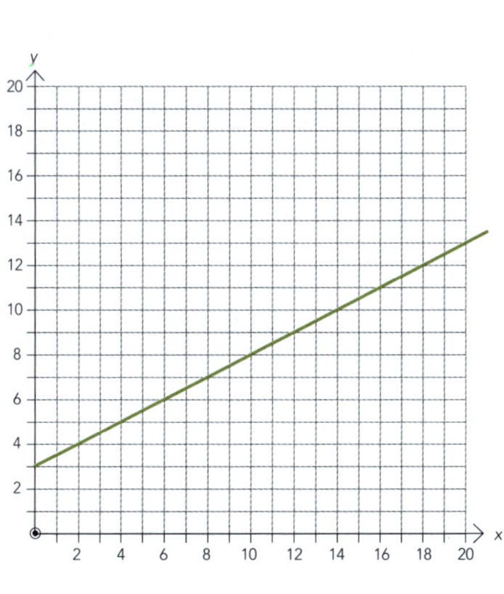

$y =$ ____________________

6

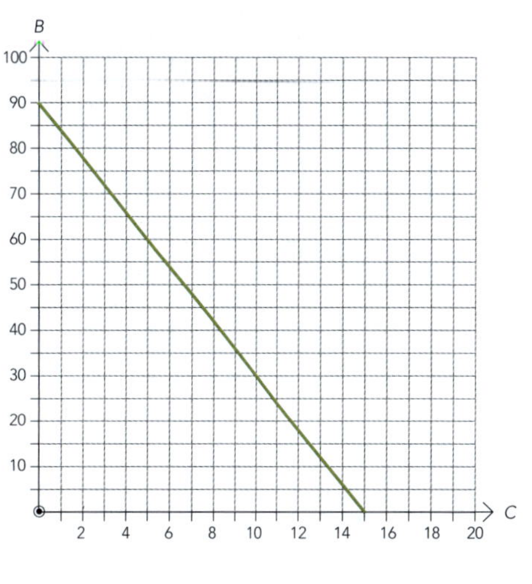

$y =$ ____________________

ISBN: 9780170370431

7

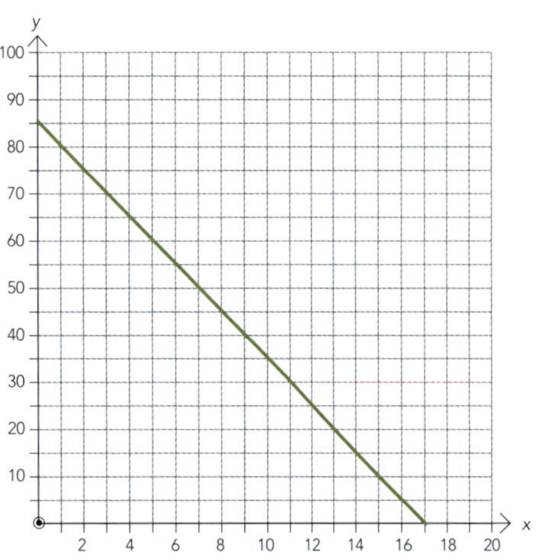

$y =$ ____________________

8

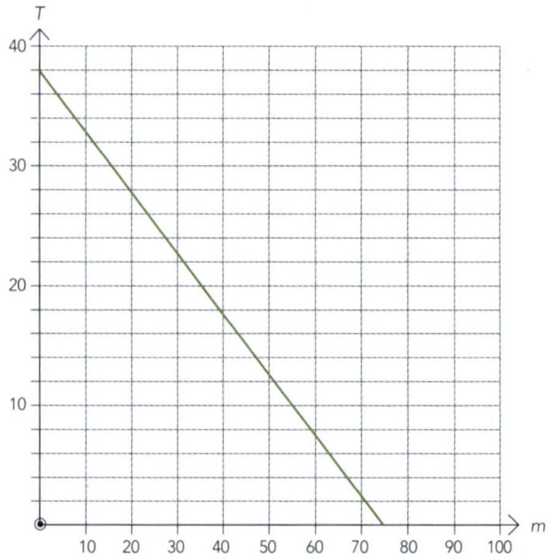

$y =$ ____________________

9

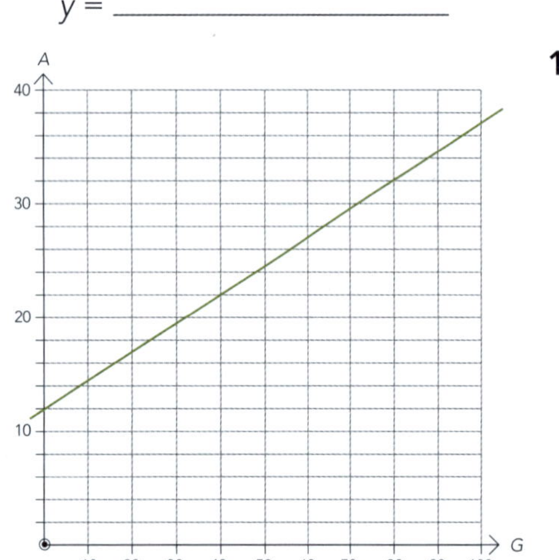

$y =$ ____________________

10

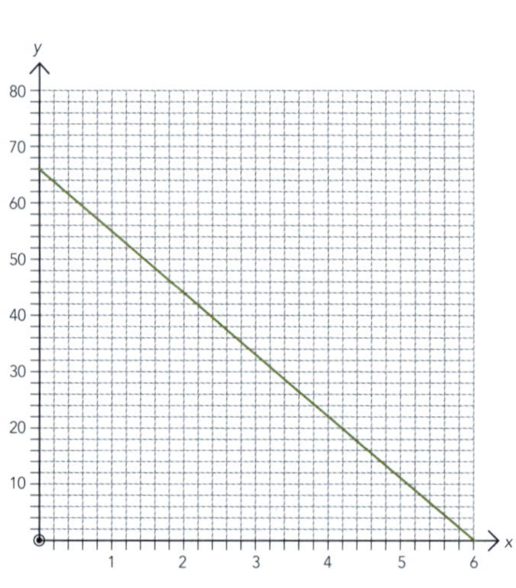

$y =$ ____________________

11

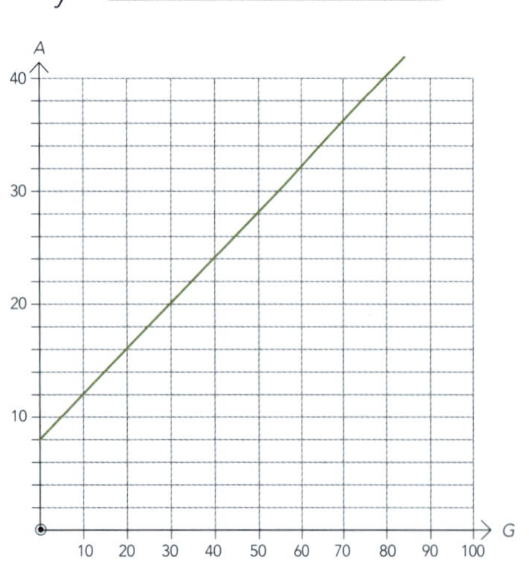

$y =$ ____________________

12

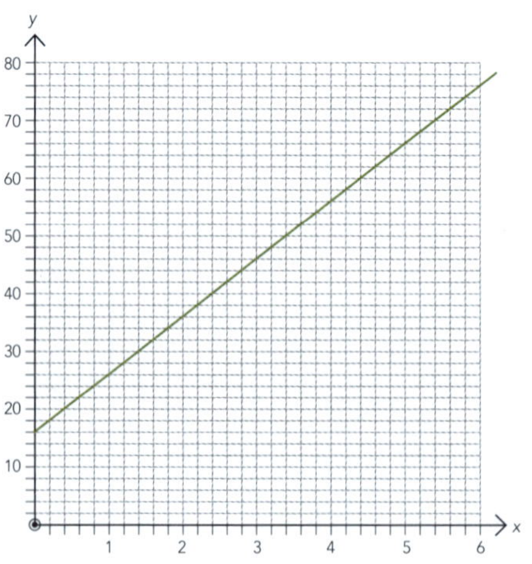

$y =$ ____________________

ISBN: 9780170370431

2 Using two points

Label the two points ($\mathbf{x_1}$, $\mathbf{y_1}$) and ($\mathbf{x_2}$, $\mathbf{y_2}$), then:

1 Substitute the co-ordinates into $\mathbf{m} = \frac{\text{rise}}{\text{run}} = \frac{y_2 - y_1}{x_2 - x_1}$ in order to find the gradient.

2 Substitute the gradient and the co-ordinates into $\mathbf{y - y_1 = m(x - x_1)}$ in order to find the equation.

Example 1:
Find the equation of the line that passes through (5, 1) and (6, 4).

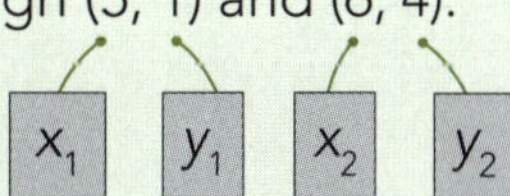

Step 1: Find the gradient.

$$m = \frac{y_2 - y_1}{x_2 - x_1} = \frac{4 - 1}{6 - 5} = \frac{3}{1} = 3$$

Step 2: Substitute this, along with x_1 and y_1, into $\mathbf{y - y_1 = m(x - x_1)}$.

$$\therefore y - 1 = 3(x - 5)$$
$$y - 1 = 3x - 15 \qquad \textbf{(+ 1)}$$
$$y = 3x - 14$$

Example 2:
Find the equation of the line that passes through (4, -3) and (-6, 2).

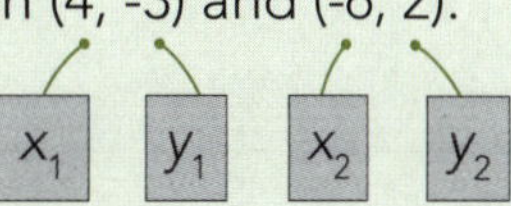

Step 1: Find the gradient.

$$m = \frac{y_2 - y_1}{x_2 - x_1} = \frac{2 - (-3)}{-6 - 4} = -\frac{1}{2}$$

Step 2: Substitute this, along with x_1 and y_1, into $\mathbf{y - y_1 = m(x - x_1)}$.

$$\therefore y - (-3) = -\frac{1}{2}(x - 4)$$
$$y + 3 = -\frac{1}{2}x + 2 \qquad \textbf{(– 3)}$$
$$y = -\frac{1}{2}x - 1$$

ISBN: 9780170370431

Calculate the equation of the line that connects the following pairs of points.

1 (1, 5) and (3, 9)

2 (2, 8) and (4, 9)

3 (5, 1) and (9, 5)

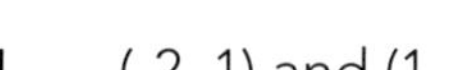

4 (-2, 1) and (1, 7)

5 (0, 3) and (1, 1)

6 (8, -2) and (1, 5)

7 (-4, 3) and (6, -2)

8 (1, -7) and (-2, 5)

ISBN: 9780170370431

9 (-3, 9) and (3, 7)

10 (0, -5) and (3, -3)

11 (1, -1) and (2, 10)

12 (4, -5) and (2, 19)

13 (1, -19) and (-7, -3)

14 (10, 9) and (-15, 19)

15 (-4, -17) and (4, -11)

16 (-4, -12) and (-16, -9)

ISBN: 9780170370431

Applications

Combine the skills that you have learnt so far in order to answer these questions.

1 Nick and Tipene are saving to go to a kapa haka competition, which will cost them $660 each. The graph shows the total amount (S) that Nick saves.

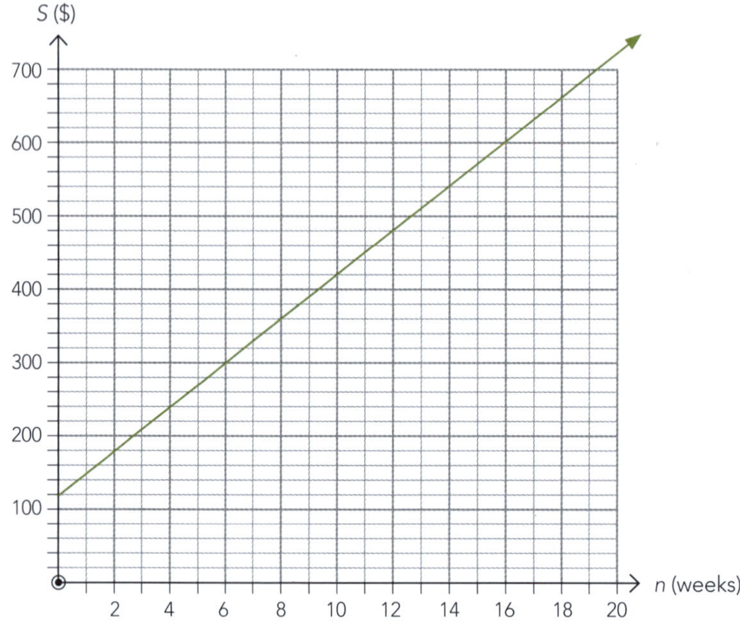

a How much did Nick have to start with? __________

How this is shown on the graph?__________

b How much does Nick save each week? __________

Show how you got your answer. __________

c Write an equation showing the relationship between the total amount Nick has saved (S) and time (n).

____

d After how many weeks will Nick have the $660 that he needs? __________
Show how you found this on the graph.

e Tipene started by cleaning some teachers' cars, and this earned him $60. He has an after-school job, so he can add $40 to his savings each week.
On the axes above, plot the graph to show how much he saves.

f Write an equation showing the relationship between the total amount Tipene has saved (S) and time (n).

____

g After how many weeks will the two boys have saved the same amount? __________

Explain how this is shown on the graph. __________

____

How much has each of them saved at this point? __________

ISBN: 9780170370431

2 Nick and Tipene's teacher needs to hire a van for the kapa haka competition. She thinks that they will drive less than 700 km.
They are looking at two companies. The graph shows the total amount (C) in dollars that Vicky's Vans will charge. Both companies charge a fixed amount, to which they add a 'per kilometre' rate.

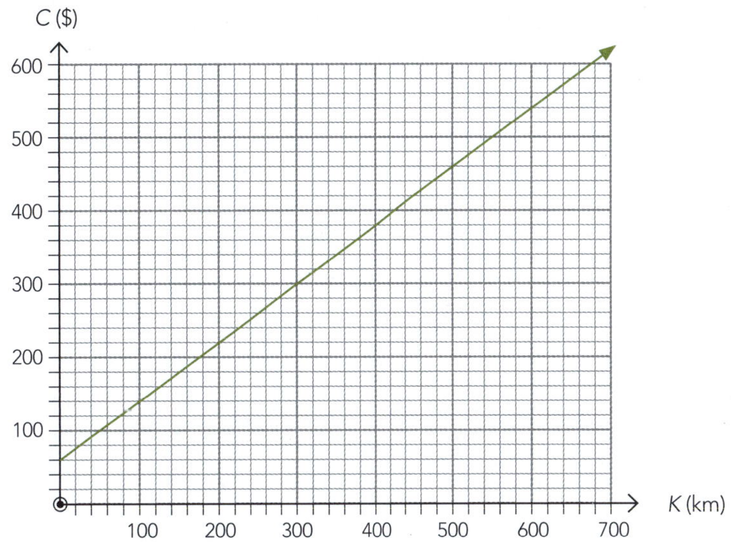

a What is the fixed charge for Vicky's Vans? ____________________

How this is shown on the graph? ____________________

b How much is the 'per kilometre' rate for Vicky's Vans? ____________________

Show how you got your answer. ____________________

c Write an equation showing the relationship between the cost of hiring Vicky's Vans (C) and the distance driven (K).

____

d How much does it cost for the drive if they drive 200 km? ____________________
Show how you found this on the graph.

e How far can they drive for $500? ____________________
Show how you found this on the graph.

f Rip-off Rentals will supply a van for the trip. It will cost a set fee of $120, plus 60c per kilometre. Draw the graph showing the cost of hiring a van from Rip-off Rentals.

g For what mileage will vans from both companies cost the same amount? ____________________

Explain how this is shown on the graph. ____________________

____

How much does it cost for this mileage? ____________________

h Write an equation for the relationship between the total cost of hiring a van from Rip-off Rentals (C) and the distance driven (K).

____

i Which company would you suggest they use? Explain your reasoning.

____

____

ISBN: 9780170370431

3 Ben borrowed some money from his dad so that he could buy himself a phone. He earns money from mowing lawns and from babysitting. He uses some of this to repay his dad equal amounts each week. The amount he owes his dad is shown on the graph.

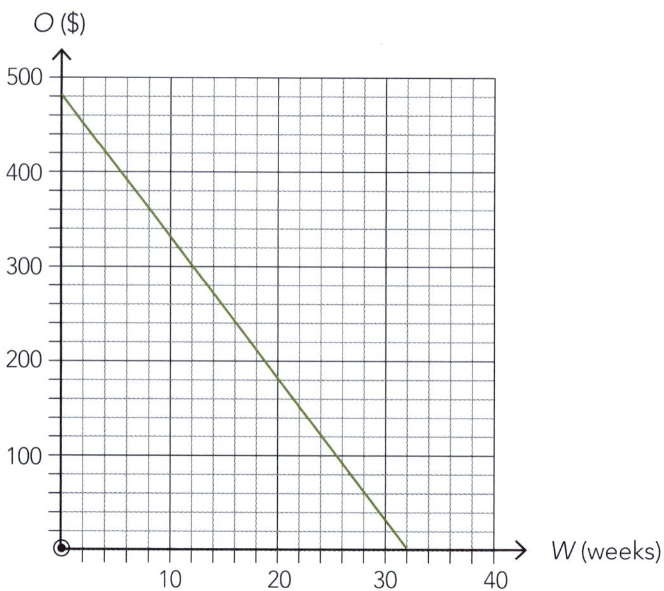

a How much did he borrow from his dad? ______

How this is shown on the graph? ______

b How much did he pay back to his dad each week? ______

Show how you got your answer. ______

c Write an equation relating the amount Ben owes his dad (O) to time (W).

______

d How much did he owe after 24 weeks? ______

Show how you found this on the graph.

e How many weeks does it take him to repay his dad? ______

Explain how this is shown on the graph. ______

______

f His father suggests that $12 might have been a better amount for Ben to repay each week. Draw a line on the graph to show this.

g If he had paid back $12 per week, how much longer would it have taken him to repay his father completely?

______

h Write an equation relating the amount he would have owed his dad if he had repaid at $12 per week ($O$) and time ($W$).

______

  ISBN: 9780170370431

4 Bridget is a full-time student. During the holidays she lived at home so she was able to save. She wants to go flatting and needs to investigate how long she can live on her savings. The graph shows the amount that she will have in the bank over the next 20 weeks.

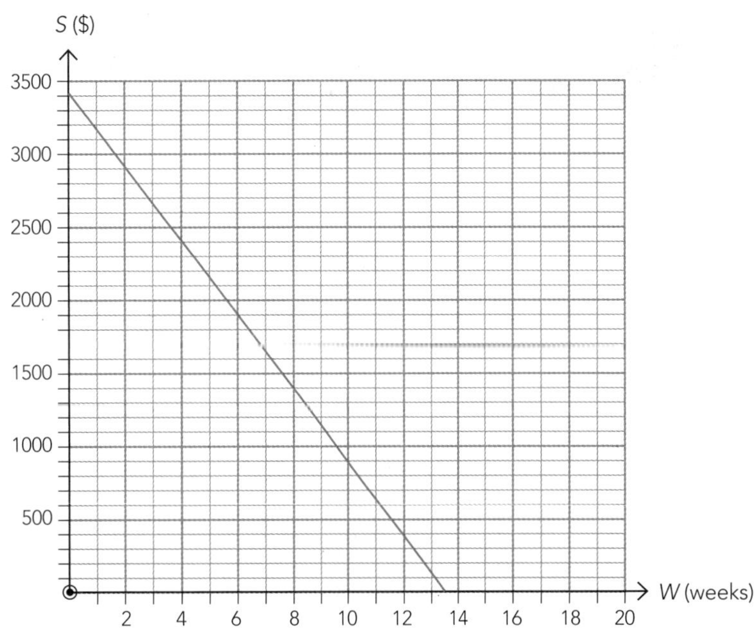

a How much does she start with? __________

How this is shown on the graph? __________

b How much has she budgeted on per week? __________

Show how you got your answer. __________

c Write an equation relating the amount (S) she has in the bank with the amount she has budgeted on each week (W).

____

d How much will she have left by the end of week 8? __________
Show how you found this on the graph.

e When will she have $400 left? __________
Show how you found this on the graph.

f She can see that her money is not going to last as long as she would like. She thinks that if she got a part-time job she could reduce the amount she draws from her savings to $150 per week. Draw the graph showing how her saving will reduce if she does this.

g Write an equation to show this new relationship.

____

h Calculate how long the $3400 will last it she reduces her spending to $150 per week.

____

i Find the difference between the amount she has left under the two plans at the end of 10 weeks.

____

ISBN: 9780170370431

Inequations

Inequations have $<$, $\leq$, $>$ or $\geq$ signs.

You solve an inequation in **exactly** the same way as you solve an equation **except**: if you need to **multiply or divide** the equation by a **negative number**, you must **reverse the sign**.

Examples:

1

$$7x < 28 \quad \textbf{(÷ 7)}$$
$$x < \frac{28}{7}$$
$$x < 4$$

2

$$5x + 3 \geq 18 \quad \textbf{(– 3)}$$
$$5x \geq 15 \quad \textbf{(÷ 5)}$$
$$x \geq \frac{15}{5}$$
$$x \geq 3$$

3

$$5 - 9x \leq 41 \quad \textbf{(– 5)}$$
$$-9x \leq 36 \quad \textbf{(÷ by -9)}$$
$$x \geq -4$$

Dividing by a **negative** number ⇒ **reverse** the sign.

4

$$\frac{17 - x}{2} > \frac{5x - 7}{3} \quad \textbf{(x by 6)}$$
$$3(17 - x) > 2(5x - 7)$$
$$51 - 3x > 10x - 14 \quad \textbf{(– 51)}$$
$$-3x > 10x - 65 \quad \textbf{(– 10x)}$$
$$-13x > -65 \quad \textbf{(÷ by -13)}$$
$$x < 5$$

Solve the following.

1 $7x \geq 21$

2 $2x - 11 < 13$

3 $19 - 6x \leq 31$

4 $4x - 7 > 6x + 3$

5 $3(5 - 4x) > -9$

6 $3(2x - 5) \geq 7(4 + x)$

7 $2 - \frac{5x}{7} > 3$

8 $\frac{4 - 3x}{5} < \frac{1 - 2x}{3}$

ISBN: 9780170370431

Forming and solving inequations

Your answers MUST have one of the following signs: $>$, $<$, $\geq$ or $\leq$, NOT an = sign.

Write and solve inequations for each of the following situations.

1 Xavier writes a puzzle. Seventeen minus five times a mystery number must be greater than two. Write an inequation and solve it to find what values the mystery number can take.

2 The length of a rectangular cake is two and a half times its width. The maximum length of a decoration which will wrap around the perimeter of the cake is 112 cm. Write an inequation and solve it to find the maximum dimensions of the cake.

3 Meg, Alannah and Suzie went out for dinner again. They spent less than $66 altogether. This time Meg's meal cost three-quarters of Alannah's, but Suzie's meal cost $4 less than Meg's. Write an inequation for the total cost of the dinner in terms of *A*, the cost of Alannah's meal. Use it to calculate the maximum costs for each meal.

4 Beth has a budget of $95 for her birthday party. She needs at least $48 for food and will use the remainder for movie tickets for herself and her friends. These cost $7.50 each. Write an inequation and solve it to find the maximum number of friends that she can invite.

5 Pete has at least $6.50 plus $\frac{3}{5}$ of the amount of money that Sam has. Write an expression for the amount Pete has (p) in terms of how much Sam has (s). If Sam has $13.50, use your expression to calculate the minimum amount Pete has.

ISBN: 9780170370431

Graphing inequations

- Lines: If the signs are ≥ or ≤, then draw a solid line: ________
  If the signs are > or <, then draw a dashed line: ....
- Shading: If you **can** have solutions in an area, leave it **blank**.
  If you **cannot** have solutions in an area, **shade it out**.

Examples:

1 $x \geq 2$

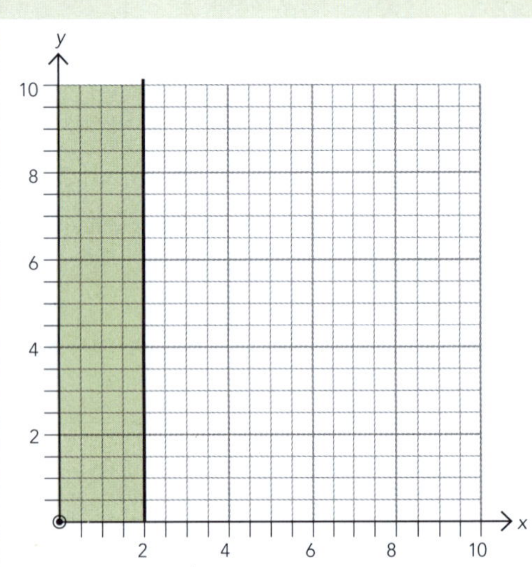

2 $y < 7$

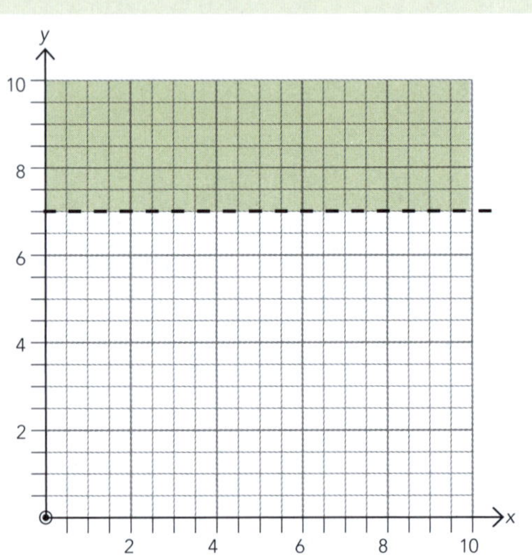

- For y **≤** or **<** something, you can have solutions in the area **below** the line, so shade out the area above it.

Examples:

1 $y \leq x + 2$

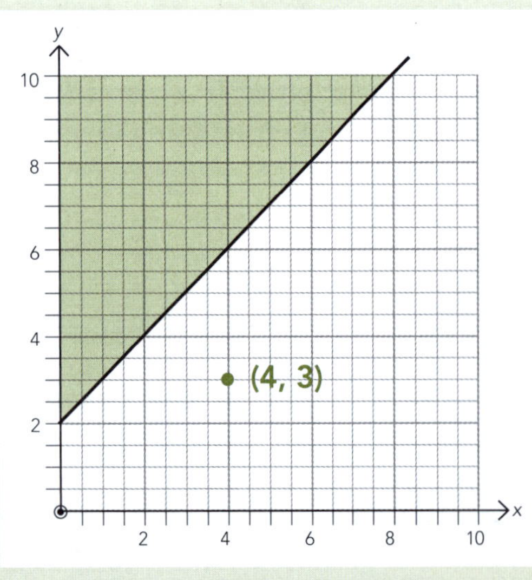

2 $y < 2x + 3$

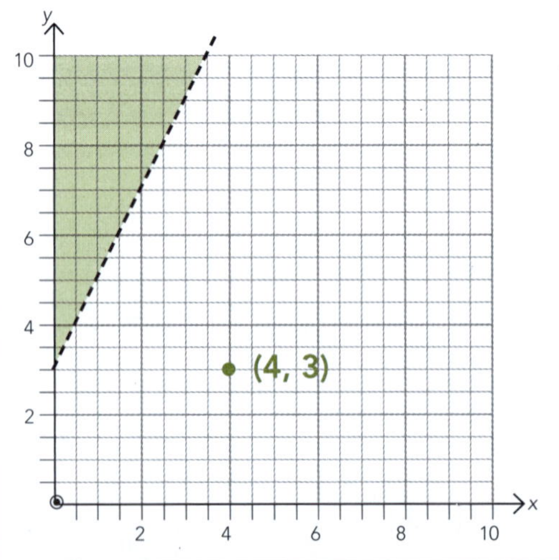

Check: The point (4, 3) would be in both these areas:

$3 \leq 4 + 2$
$3 \leq 6$ ✓

$3 < 2(4) + 3$
$3 < 11$ ✓

ISBN: 9780170370431

- For $y \geq$ or $>$ something, you can have solutions in the area **above** the line, so shade out the area below it.

Examples:

1 $y \geq x + 2$

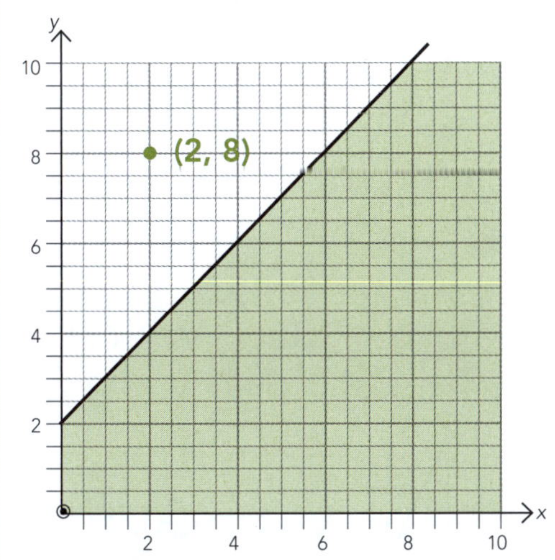

2 $y > 2x + 3$

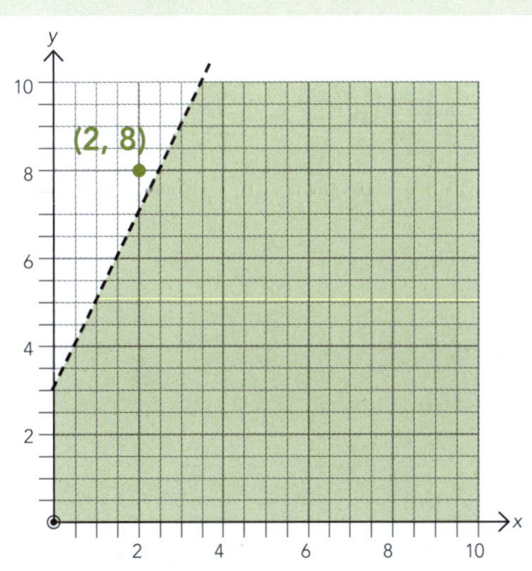

Check: The point (2, 8) would be in both these areas:

$8 \geq 2 + 2$
$8 \geq 4$ ✓

$8 > 2(2) + 3$
$8 > 7$ ✓

Select the correct inequation for each of these graphs.

1

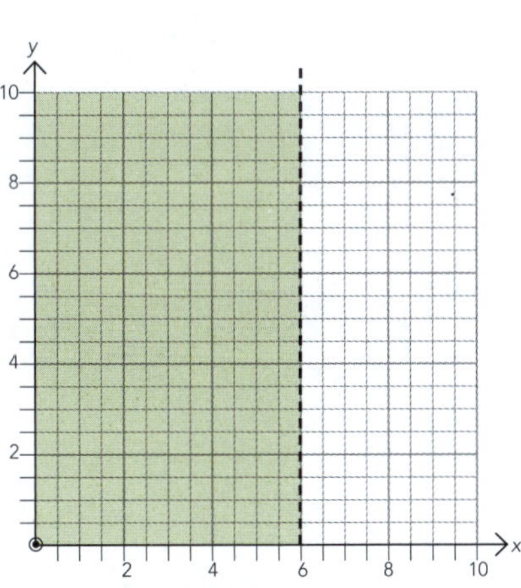

$x > 6$ $y > 6$
$x \geq 6$ $y \geq 6$
$x < 6$ $y < 6$
$x \leq 6$ $y \leq 6$

2

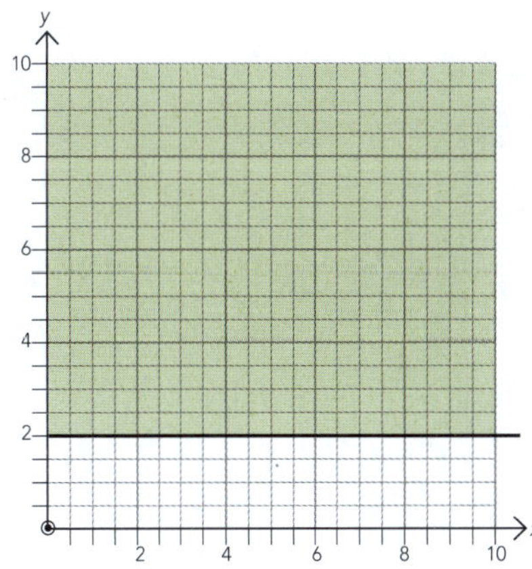

$x > 2$ $y > 2$
$x \geq 2$ $y \geq 2$
$x < 2$ $y < 2$
$x \leq 2$ $y \leq 2$

ISBN: 9780170370431

3

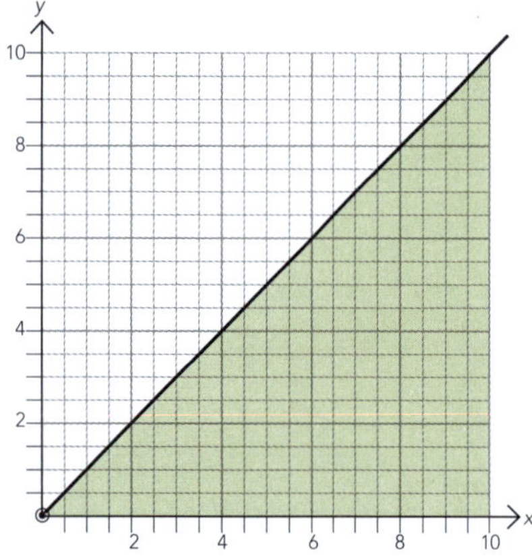

$y > x$
$y \geq x$
$y < x$
$y \leq x$

4

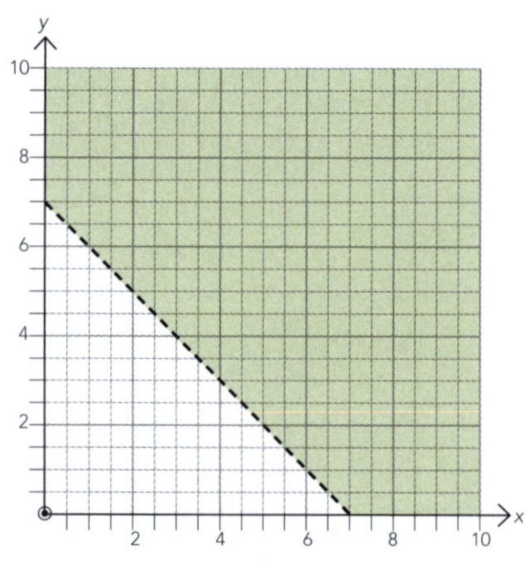

$y > -x + 7$
$y \geq -x + 7$
$y < -x + 7$
$y \leq -x + 7$

5

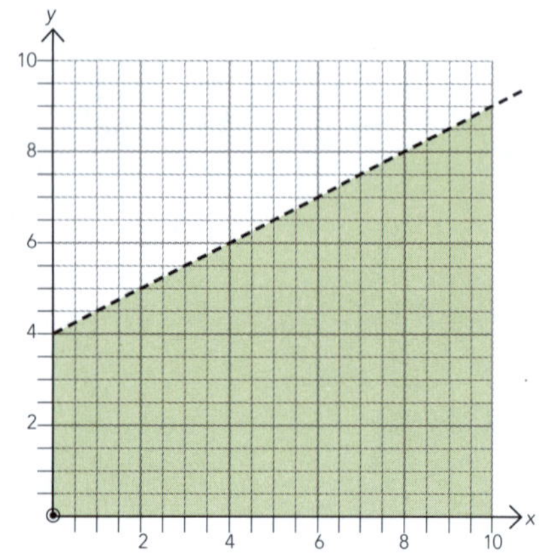

$y > \frac{1}{2}x + 4$ $\quad$ $y \geq \frac{1}{2}x + 4$

$y < \frac{1}{2}x + 4$ $\quad$ $y \leq \frac{1}{2}x + 4$

6

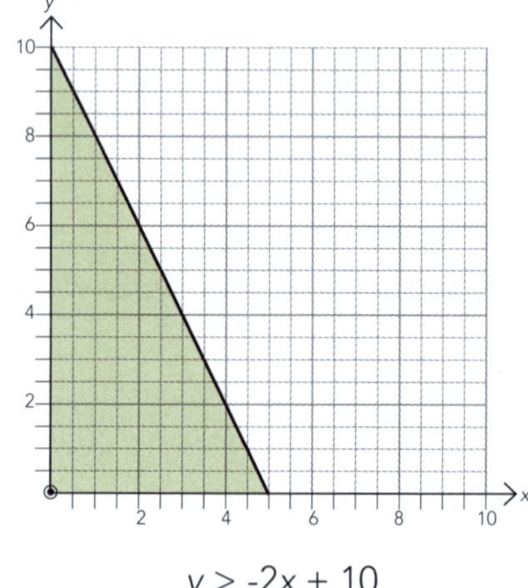

$y > -2x + 10$
$y \geq -2x + 10$
$y < -2x + 10$
$y \leq -2x + 10$

7

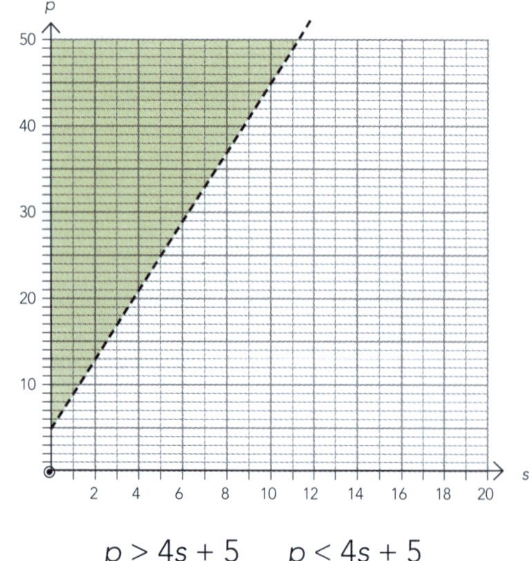

$p > 4s + 5$ $\quad$ $p < 4s + 5$
$s > 4p + 5$ $\quad$ $s < 4p + 5$

8

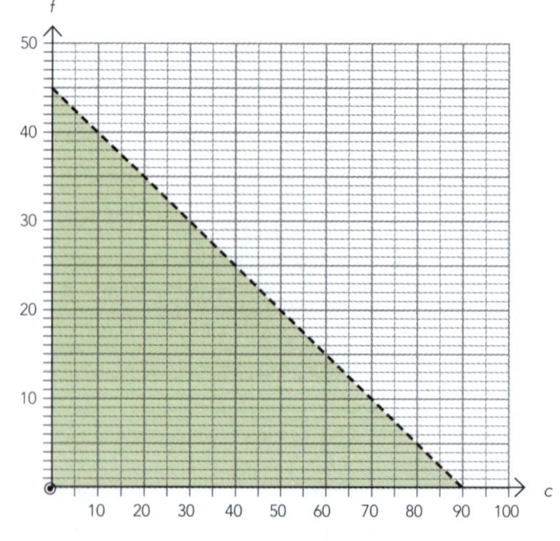

$f \geq -0.5c + 45$ $\quad$ $c \geq -0.5f + 45$
$f \leq -0.5c + 45$ $\quad$ $c \leq -0.5f + 45$

ISBN: 9780170370431

Draw graphs to match the following inequations.

9  $y > 4$

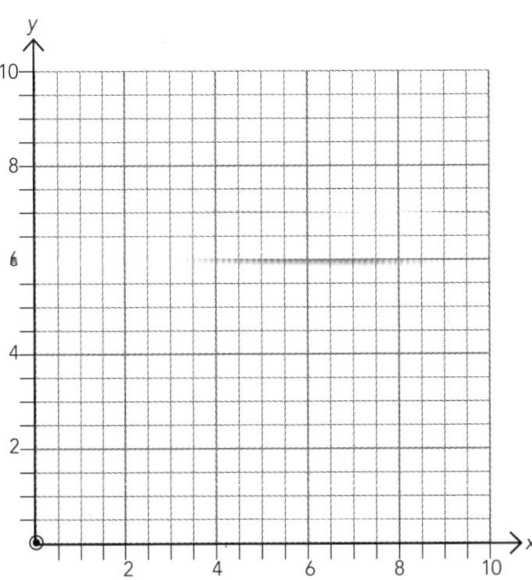

10 $x \leq 9$

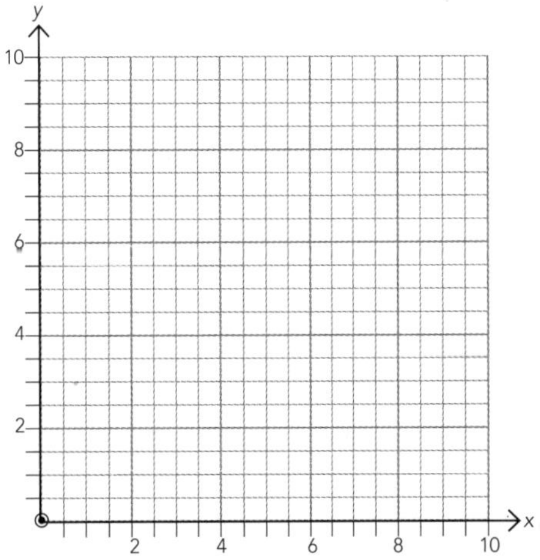

11 $y < x + 1$

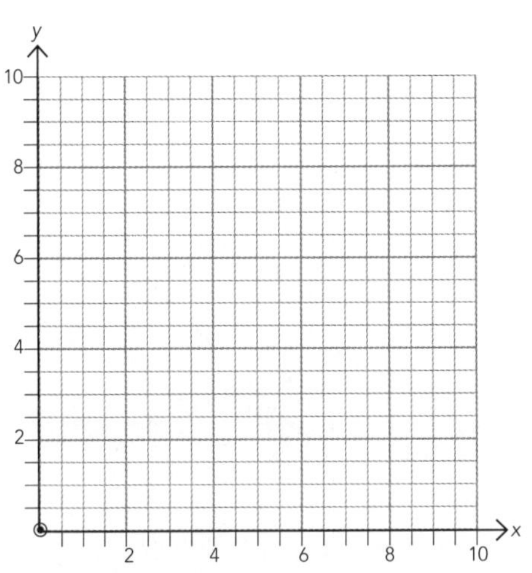

12 $y \geq -x + 5$

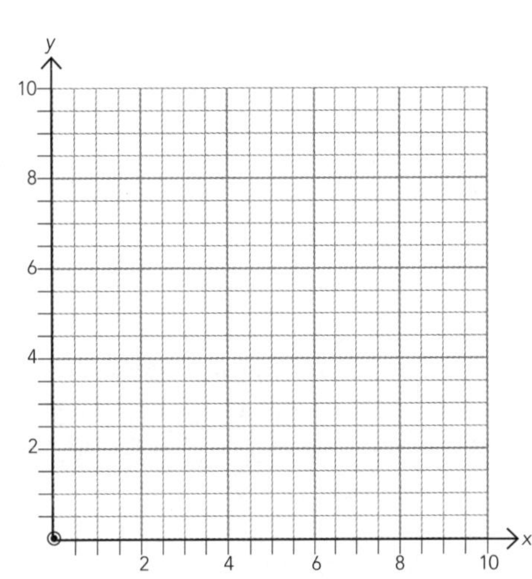

13 $y > -\frac{3}{2}x + 9$

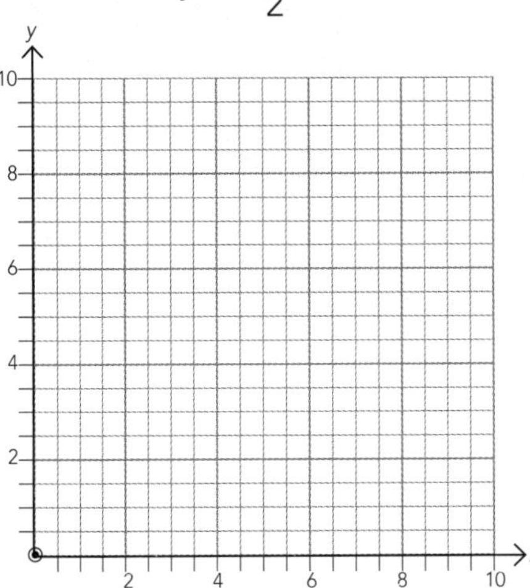

14 $y \leq 2x$

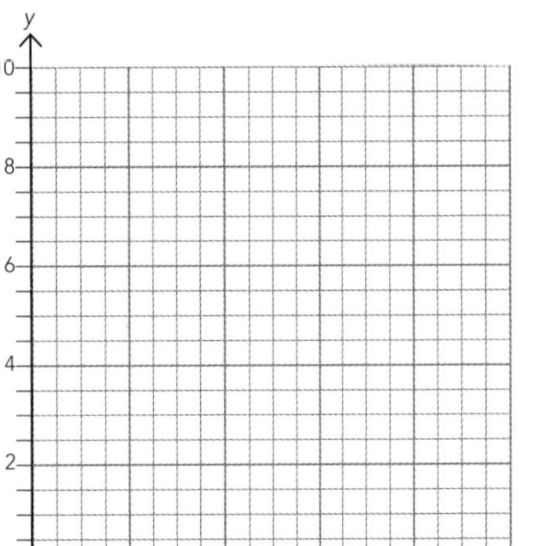

ISBN: 9780170370431

Combining several inequations

Meanings of expressions:

more than, over	$>$
less than, under, fewer than	$<$
at least, a minimum of, no less than, greater than or equal to	$\geq$
a maximum of, no more than, less than or equal to	$\leq$

Example 1: Fred has fewer than twice the number of lollies that Tomas has. If Tomas had four more, Fred would still have at least as many as Tomas. What is the minimum number of lollies that each could have?

Step 1: Define your variables.
Let t represent the number of lollies that Tomas has.
Let f represent the number of lollies that Fred has.

Step 2: Write equations or inequations to represent the relationships.

$$f < 2t$$
$$f \geq t + 4$$

Step 3: Plot the equations or inequations.

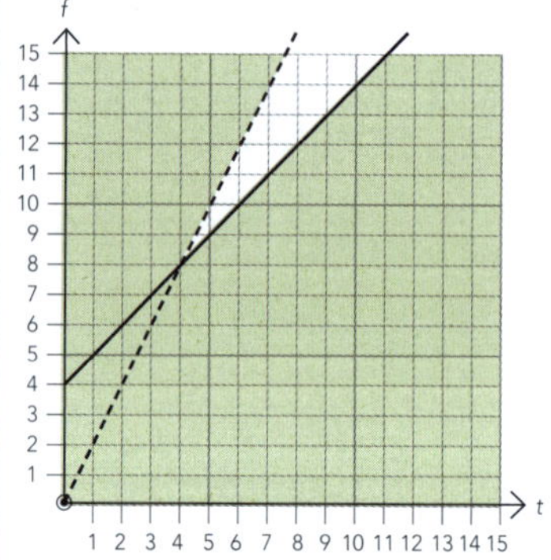

Step 4: Possible solutions
- can lie only in the unshaded area or along a solid line on its border
- must be whole numbers of lollies.

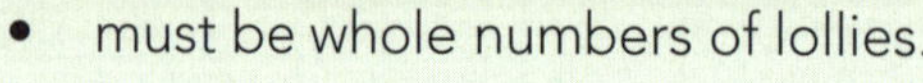

Step 5: Show these on the graph.

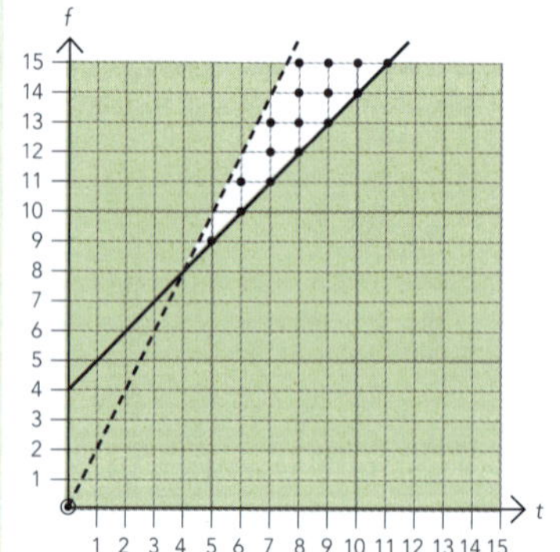

You are asked for the smallest number of lollies each can have, so (5, 9) gives the solution.
Check: $f < 2t \Rightarrow 9 < 10$ ✓
and $f \geq t + 4 \Rightarrow 9 \geq 5 + 4$ ✓
Note: (4, 8) is not a solution because Fred has to have **fewer than** twice the number as Tomas.

Step 6: Write your answer in a sentence: *Tomas has five lollies and Fred has nine.*

ISBN: 9780170370431

Example 2: Toby and Georgia both own taxis. Both charge fares to the nearest kilometre.
Toby charges no more than a \$15 base fee plus \$2 per kilometre (in total).
Georgia always charges \$5 per kilometre.
Show these fares on a graph. Use the graph to find the cheapest fares for distances up to 15 km. Give examples in your answer.

Step 1: Define your variables.
Let k represent the number of kilometres travelled.
Let f represent the total price of a fare.

Step 2: Write equations or inequations to represent the relationships.

Toby: $f \leq 2k + 15$
Georgia: $f = 5k$

Step 3: Plot the equations or inequations.

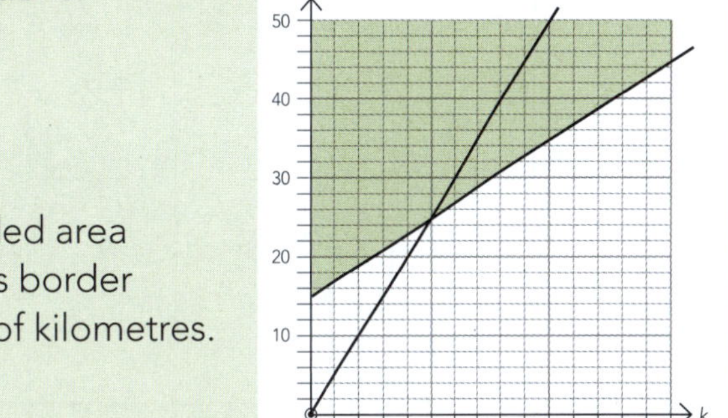

Step 4: Possible solutions

- can lie only in the unshaded area or along a solid line on its border
- must be whole numbers of kilometres.

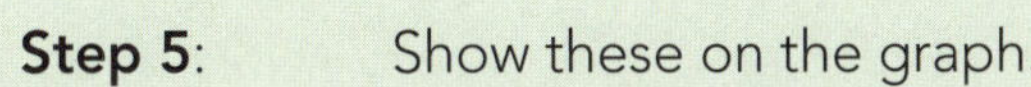

Step 5: Show these on the graph.

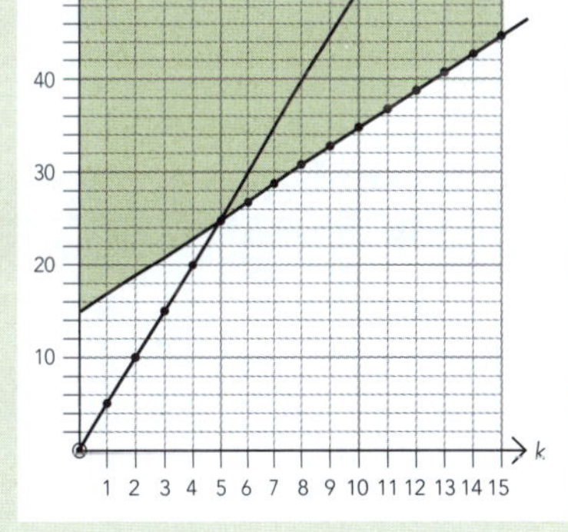

Step 6: Write your answer in a sentence.

For distances less than 5 km, Georgia's taxi is probably cheapest. We know only the maximum fare that Toby charges, so he could undercut her. Example: For a distance of 4 km, Georgia would charge \$20 and the maximum Toby would charge is \$23. If he gave a \$4 discount, he would be the cheapest.

For a journey of 5 km, both charge \$25, but once again, Toby could charge less.

For distances of more than 5 km, Toby definitely has the cheapest rates. Example: For a distance of 8 km, Toby would charge a maximum of \$31 and Georgia would charge \$40.

ISBN: 9780170370431

Write equations and inequations, draw graphs and find the solutions to the following.

1 Sebastian has more than three times the number of lollies that Alexander has. If Alexander had eight more, he would have the same as Sebastian. Calculate the maximum and minimum numbers of lollies that each could have.

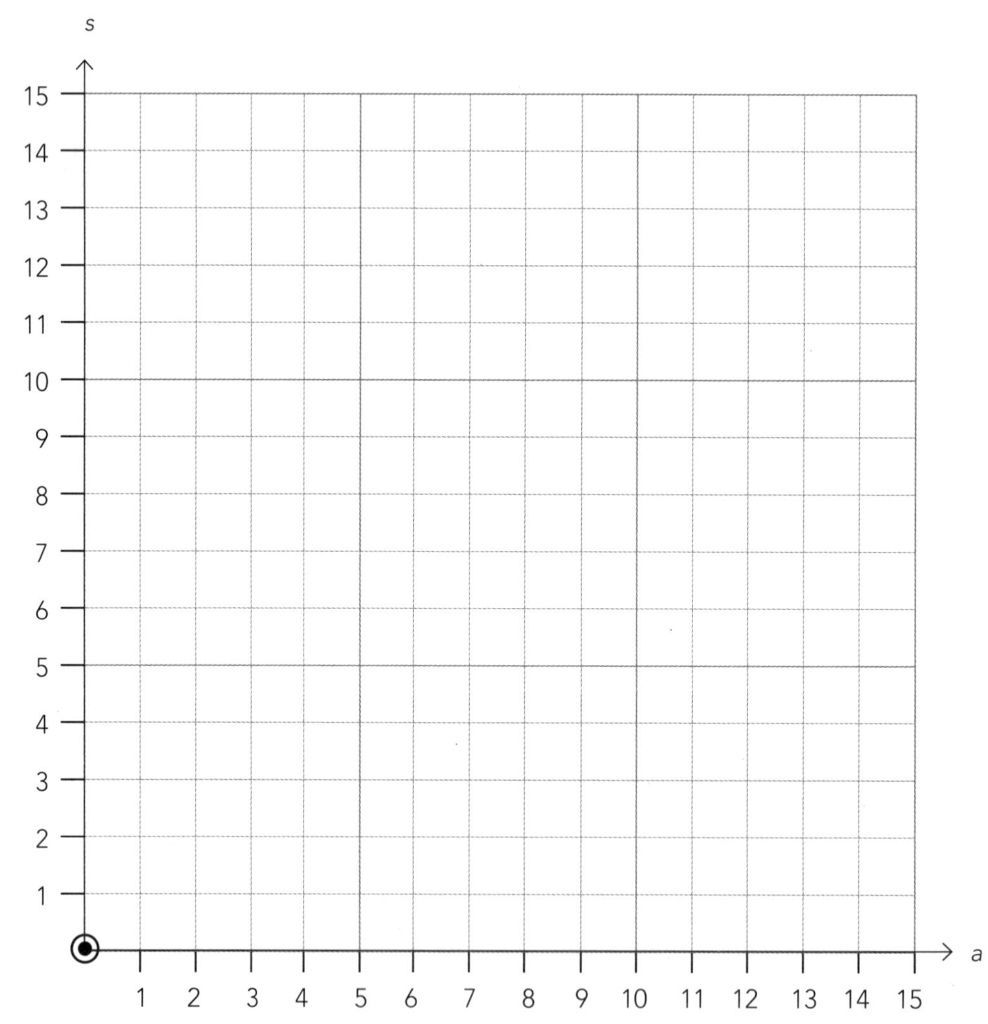

ISBN: 9780170370431

2 Florrie has at least five more than the number of lollies Rosa has. If Rosa had 16 more, she would have more than double the number that Florrie has. Calculate the maximum number of lollies that each could have.

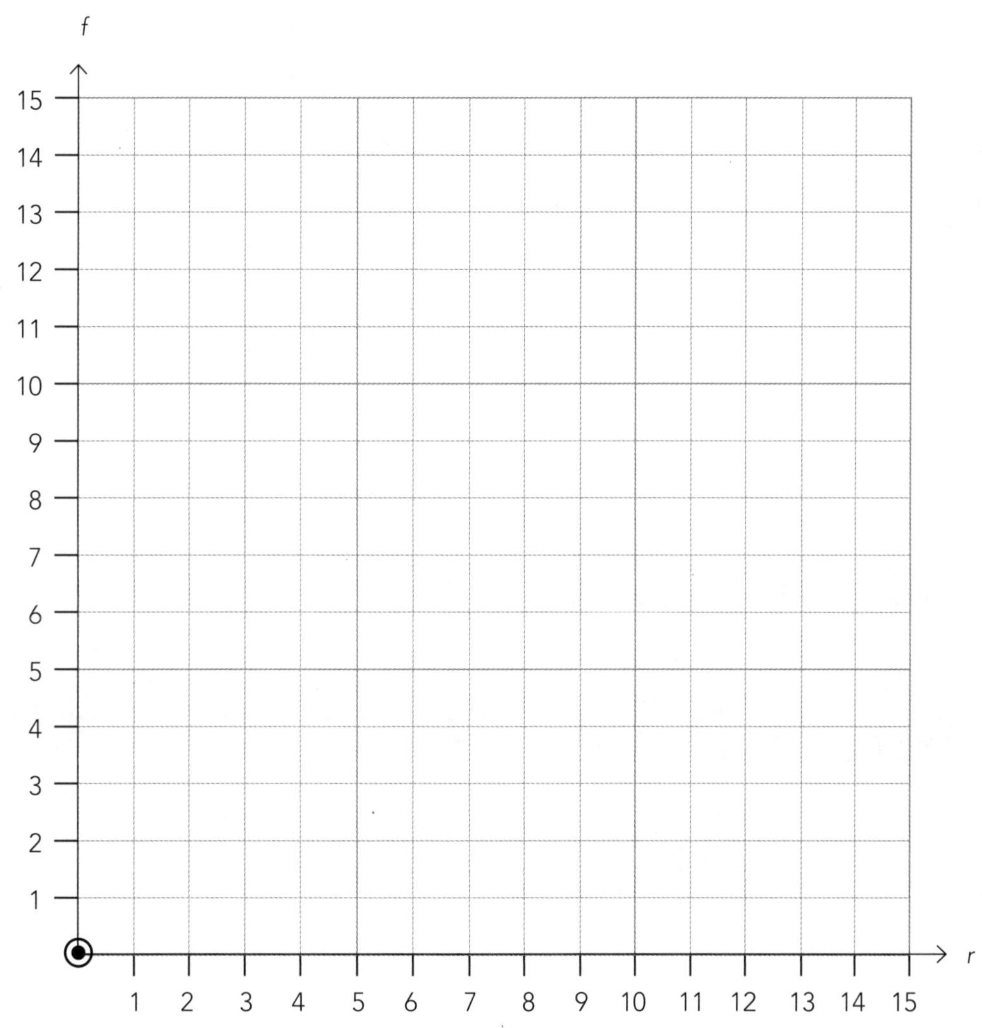

3 The school council is hiring a bouncy castle for its gala day. Bob's Bounce charges an \$80 set-up fee plus \$20 per hour. Craig's Castles charges no set-up fee and at least \$40 per hour, depending on any damage that occurs. Use your graph to find the cheapest deal for hiring a castle for up to 10 hours. Give examples of the charges.

ISBN: 9780170370431

4 The council also needs to hire a big tent. Chrissie's Canvas will charge $100 to put the tent up, plus $20 per hour. Tom's Tents will charge less than $200 for the day. Use your graph to find the cheapest deal for hiring the tent for up to 10 hours. Give examples of the charges.

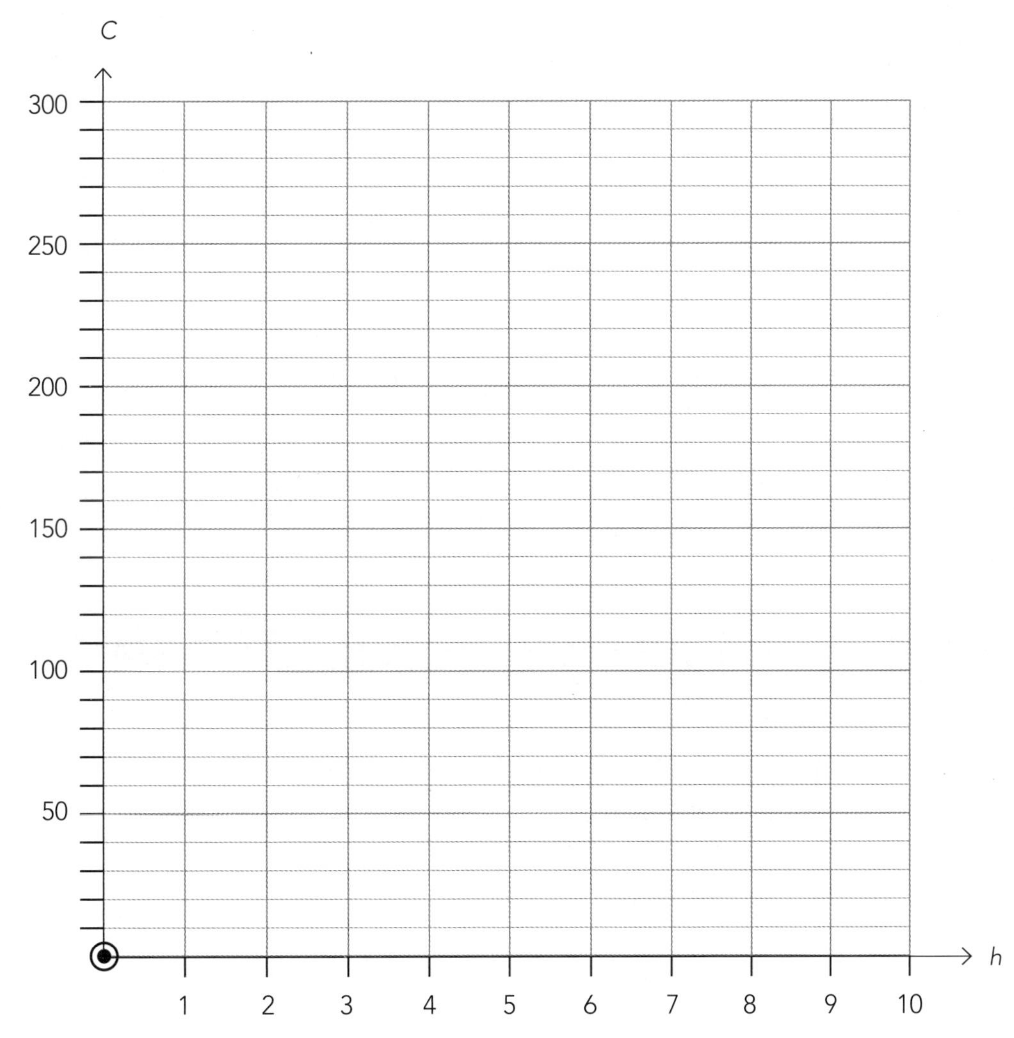

ISBN: 9780170370431

Piecewise functions

These occur when there are **different** rules for finding y, depending on the value of x.

Example: At Fergie's Fish, the cost of a whole warehou depends on how heavy it is. A warehou that is less than 1 kg costs $15. A warehou that is 1 kg or more, but less than 2 kg, costs $20. The price increases in $5 steps for each kilogram.

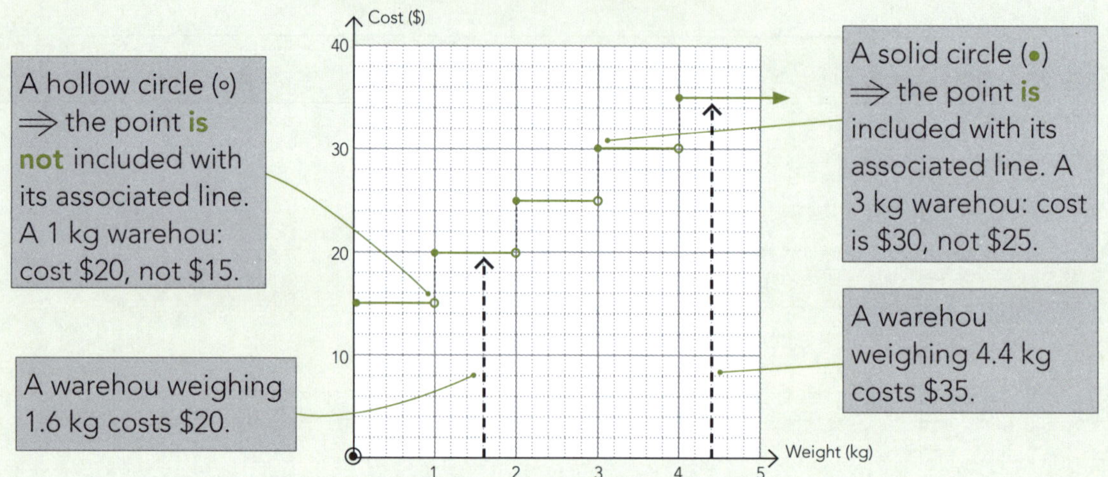

You can use the graph to work out the cost for any weight of warehou.
The scale of charges is different at Snappy Seafood. They charge $10 per kilogram.
Both rates are shown on the graph below.

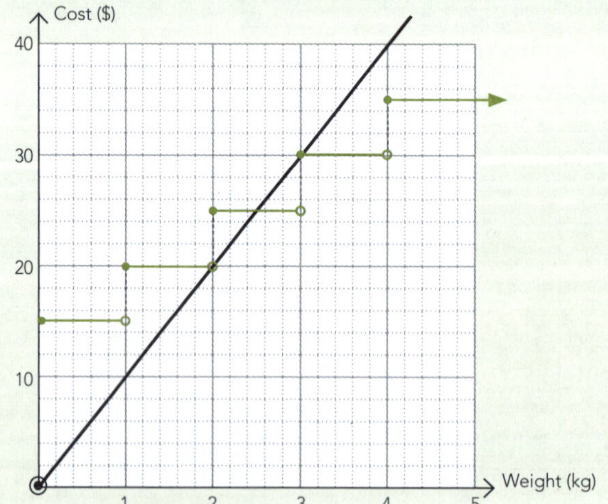

Compare the prices of buying fish at the two shops. Use examples in your comparison.

For fish that weigh less than 2.5 kg it is cheaper to buy from Snappy Seafood.
For example, a 1.6 kg fish costs $16 at Snappy Seafood, but it costs $20 at Fergie's Fish.
A fish that weighs exactly 2.5 kg or 3.0 kg costs the same at both shops. They cost $25 or $30 respectively.
For fish over 2.5 kg (apart from a 3 kg fish), it is cheaper to buy at Fergie's Fish.
For example, a 3.8 kg fish costs $30 at Fergie's Fish, but $38 at Snappy Seafood.

ISBN: 9780170370431

Draw graphs for the following relationships, and use them to answer the questions.

1 The cost of posting a parcel by air (weighed to the nearest gram) is shown in the table below.

Weight (g)	Cost ($)
0–500	18.00
501–1000	25.00
1001–2000	31.00
2001–3000	36.00

Show these charges on the graph.

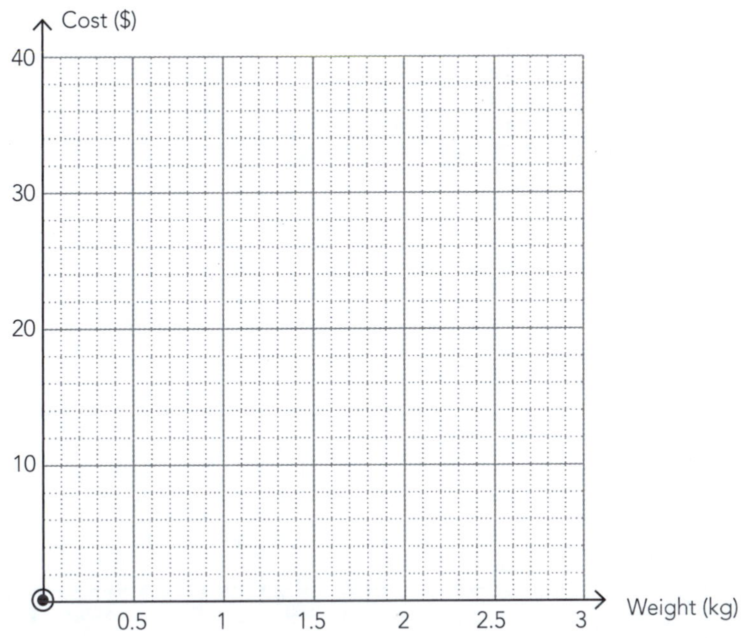

A courier company charges a set fee of $11 plus $10 per kilogram. For example, a 1.8 kg parcel will cost $(11 + 1.8 x 10) = $29. Add these charges to your graph.

Compare the charges for the two methods of transporting parcels.

ISBN: 9780170370431

2 A school is having its magazines printed. The cost of printing each magazine depends on how many are printed. The table below shows the prices quoted by Pete's Printing.

Number of copies (n)	Cost per copy ($)
0–199	5.50
200–999	4.00
1000–1999	3.00
> 2000	2.20

Show these charges on the graph.

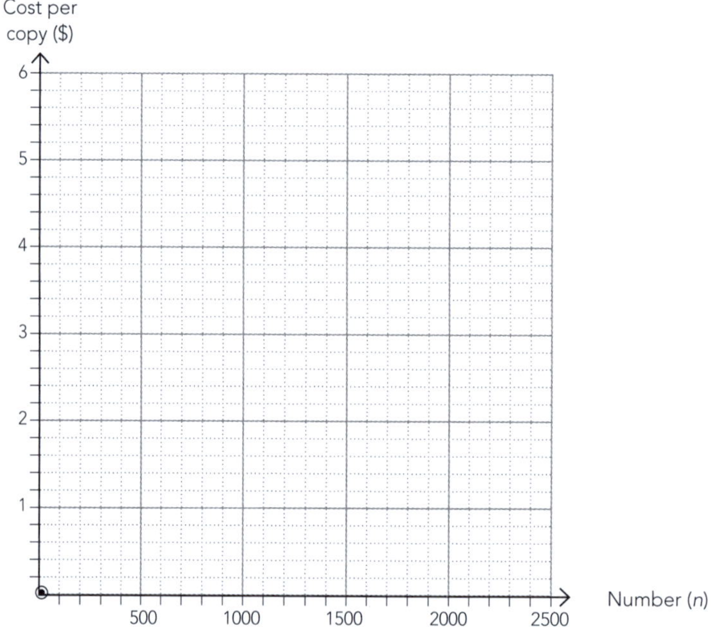

Maria's Mags' cost per magazine is $6 minus $0.002 for every extra magazine printed (e.g. if you print 2500 magazines, they cost $1.00 each). Add her charges to your graph.

Compare the charges for the two printers.

 ISBN: 9780170370431

3 Tiger Taxis charges a fixed charge of \$10 plus \$2 per kilometre for the first 10 km, and \$1.50 per kilometre for extra distance above 10 km.

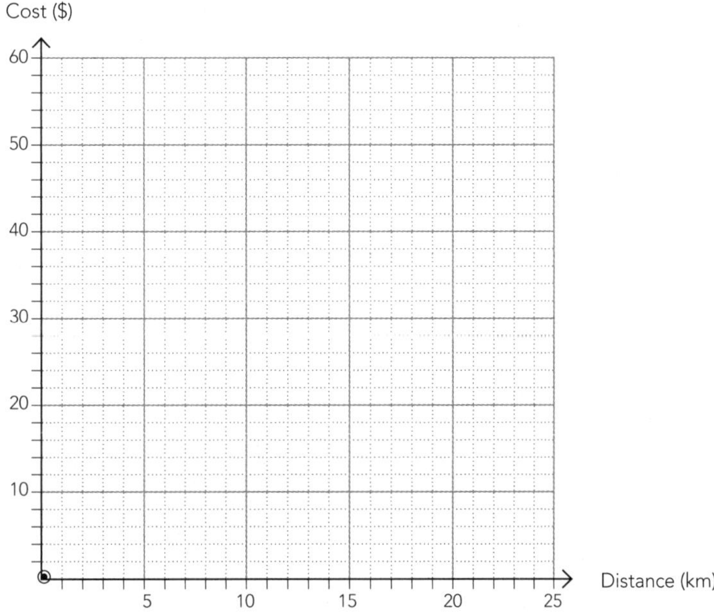

a Draw these charges on the graph.

b Find the equations for each section of the graph.

c How much do they charge for distances of

9 km?

18 km?

d Claude's Cabs don't want to do calculations, and they don't want long trips, so they just charge \$25 for any trip up to 20 km. Add this to your graph and write an equation for the line.

e Compare the prices for the two taxi companies, including examples of charges for several distances.

ISBN: 9780170370431

4 Penny the plumber charges a $30 call-out fee, plus $40 per hour for the first four hours, and $35 per hour for extra hours above four hours.

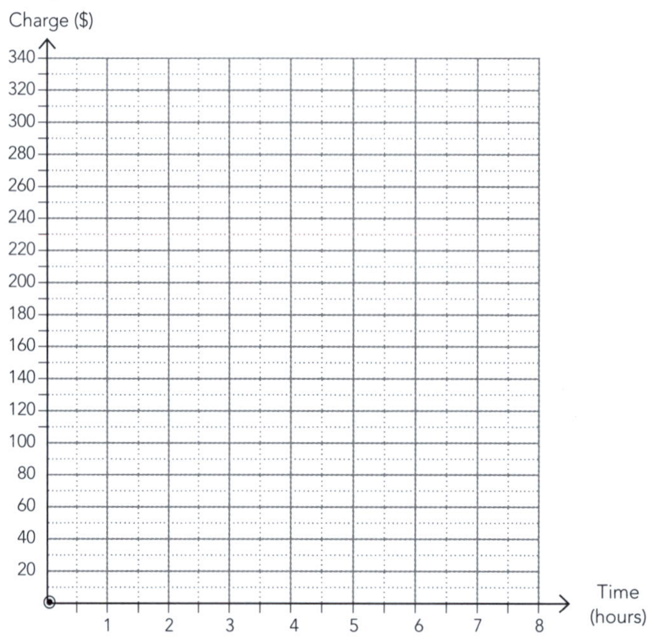

a Draw these charges on the graph.

b Find the equations for each section of the graph.

---

c How much do they charge for distances of

three hours' work? ____________________

eight hours' work? ____________________

d No Drips charge $130 for all jobs less than four hours long. For any time above that they charge $50 per hour. Draw their charges on the graph, and write down equations for each section.

---

e Compare the prices for the two plumbers, including examples of charges for several different times.

 ISBN: 9780170370431

Practice tasks

Practice task one

A popular tourist city has three bike hire companies: Ben's Bikes, Sally's Cycles and Will's Wheels. They all round their charges up to the nearest whole hour. Their charges are given below.

Ben's Bikes	**Sally's Cycles**	**Will's Wheels**
Fixed charge: \$6.00. Per hour: \$2.	$C = 0.5H + 15$ where C is the cost in dollars and H is the time in hours.	\$20.00 flat fee for a whole day.

- Represent the bike hire companies' charges using the same representation (for example, three equations or three graphs).
- Three tourists want to hire bikes (see table below). Recommend which bike hire company each should use.
- Recommend distances for which it would be cheapest to use Sally's Cycles.
- Ben, who owns Ben's Bikes, wants to be the cheapest bike hire company that tourists can use for any period. Describe at least two different ways Ben could realistically change his hire charges to achieve this goal. Include specific examples of the rates he could use.

Hire periods

Name	Hours
Hemi	2
Cleo	6
Didier	9

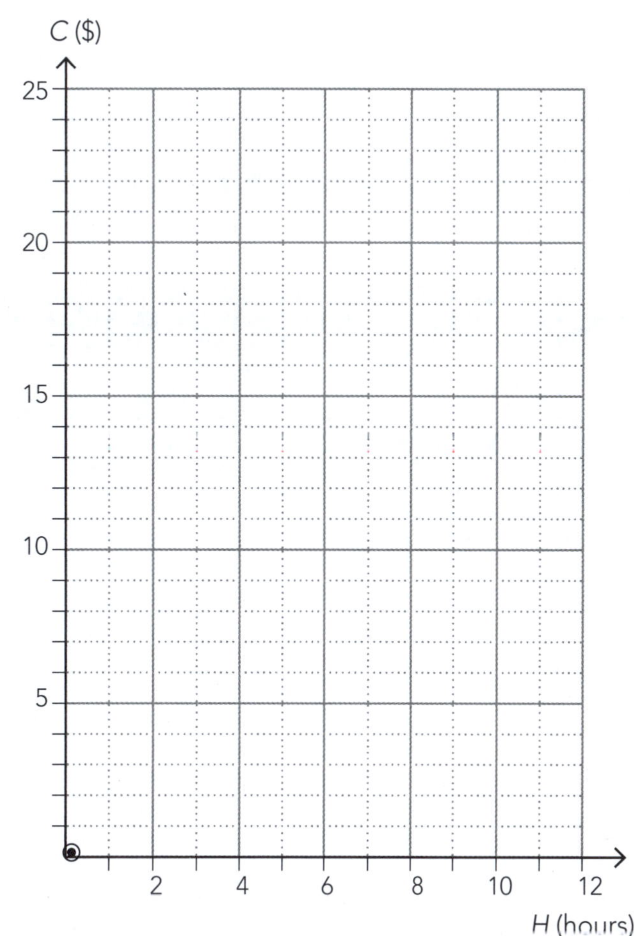

ISBN: 9780170370431

ISBN: 9780170370431

Practice task two

The school council wants to hire a candy floss machine for the day of the school fair. They investigate three companies, whose charges are given below. The council agrees that the maximum number of servings they will need is 200.

Felicity's Floss	**Pita's Party Hire**	**Colin's Candy**
Fixed charge: \$80.00. Per 100 servings: \$25.	$C = 0.5S + 50$ where C is the cost in dollars and S is the number of servings.	\$120.00 flat fee for the day and up to 200 servings.

- Represent the candy floss hire charges using the same representation (for example, three equations or three graphs).
- Recommend whose machine they should hire for a range of different numbers of servings.
- Pita wants to be the cheapest candy floss hire company, regardless of how many servings are made. Describe at least two different ways he could realistically change his hire charges to achieve this goal. Include specific examples of the rates he could use.

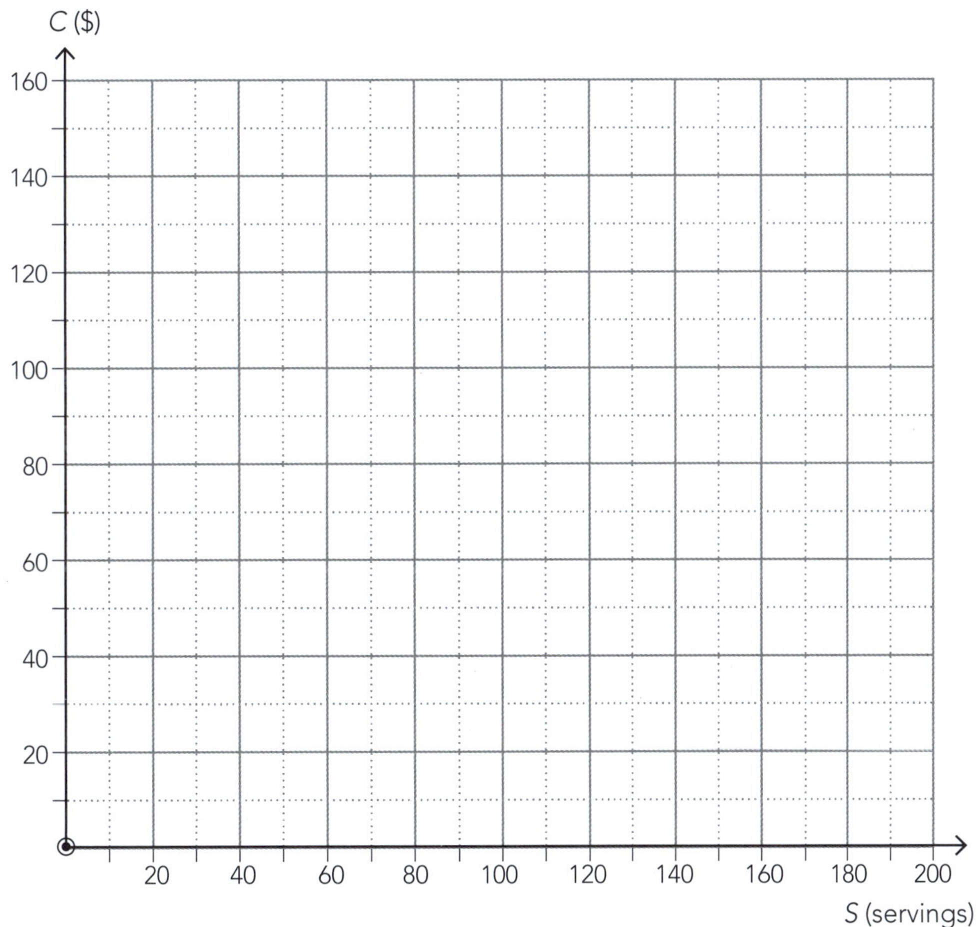

ISBN: 9780170370431

ISBN: 9780170370431

Practice task three

Marama needs to borrow $600 for a computer. Her mum, her dad and her grandma will all lend her the $600. Their repayments are shown in the table below.

Mum	**Dad**	**Grandma**
Pay $300 at the end of the first year. Pay the remaining $300 after two years.	$O = -20M + 600$ where O is how much she owes in dollars and M is the number of months.	Pay back $15 per month for the first 12 months, then $30 per month until all the money is repaid.

- Represent the amount she owes using the same representation (for example, three equations or three graphs).
- How long would she take to repay all of the $600 to each person?
- State how much she would owe each person after 12 months, 18 months and 24 months.
- Grandpa says that he would lend her the $600 and immediately reduce the amount she owes him to $500 as an advance birthday present. Suggest at least two repayment schemes that would enable her to pay her grandfather back more quickly than some or all of the other three repayment schemes.

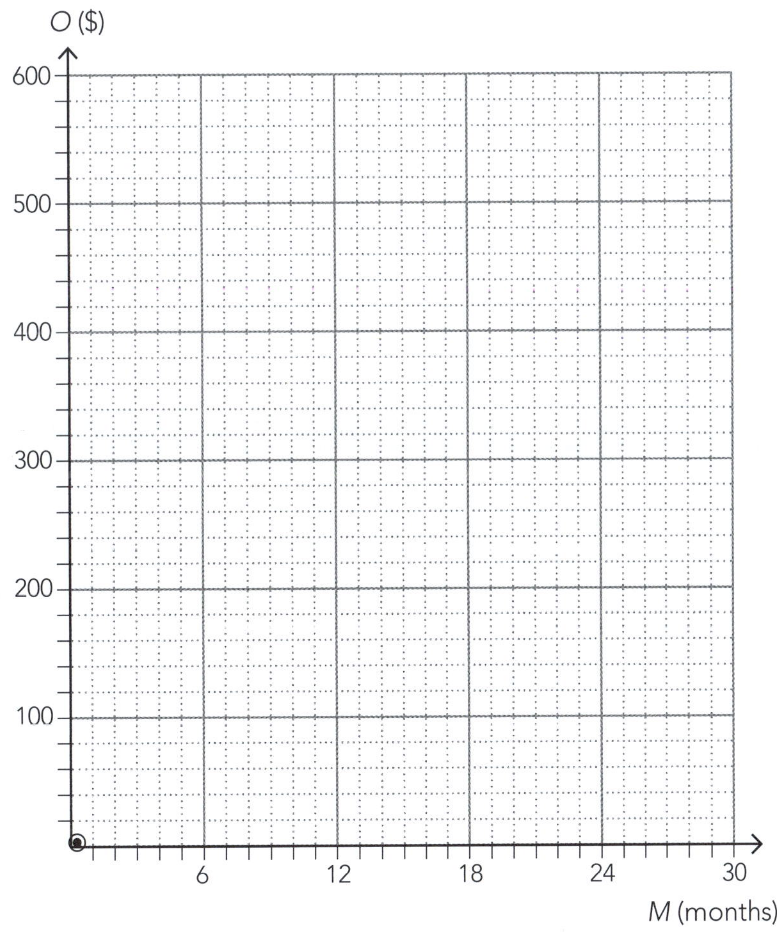

  ISBN: 9780170370431

Answers

Substitution (p. 5)

1	-15	**2**	1
3	24	**4**	-2
5	-17	**6**	-20
7	9	**8**	-5
9	4	**10**	2
11	1	**12**	7

Solving linear equations (pp. 6–14)

One-step equations (p. 6)

1	21.6	**2**	2.2
3	3.4	**4**	5
5	-88	**6**	-2.18
7	7.3	**8**	0.7
9	3.2	**10**	5.28

Two-step equations (p. 7)

1	9.2	**2**	9
3	7	**4**	16
5	-10	**6**	-0.05
7	0.6	**8**	27
9	18.6	**10**	-0.014

Equations with a variable on both sides (p. 8)

1	5	**2**	8
3	6.5	**4**	-14.5
5	-17	**6**	4.2
7	17.5	**8**	-53.3
9	19	**10**	-0.09

Equations with brackets (p. 9)

1	17	**2**	-2
3	7	**4**	-0.5
5	1	**6**	2
7	4	**8**	0
9	-1.4	**10**	1.25

Equations with fractions (p. 10)

1	12	**2**	10
3	4	**4**	2.5
5	-6	**6**	$-\frac{4}{7}$
7	$-\frac{1}{2}$	**8**	3

Forming and solving linear equations (pp. 11–14)

1 Three adults and five children.
2 Teachers \$11, students \$7.
3 Large \$5.50, small \$4.00.
4 Kowhai 5, lancewoods 4.
5 In six years' time, Siale will be 11 and her mum will be 33.
6 Six years ago, Alfie was 12 and his father was 36.
7 Albert is 48 and Simon is 16, so in 16 years' time, Albert will be 64 and Simon will be 32.
8 The rectangle is 12 cm by 9 cm.
9 The parallelogram is 15 cm by 8 cm.
10 Sides are 12 cm, 36 cm and 35 cm.
11 Meg \$27, Alannah \$18 and Suzie \$21.
12 Length = 48 cm, depth = 32 cm.
13 He gave Hine two cherries at the start.

Rearrangement of expressions (pp. 15–16)

1	$x = \frac{y-8}{5}$	**2**	$x = \frac{y+6}{3}$
3	$x = \frac{3y-12}{2}$	**4**	$x = \frac{y-25}{6}$
5	$x = 3y + 19$	**6**	$x = \frac{4y+30}{3}$
7	$x = \frac{6y-29}{2}$	**8**	$P = \frac{100\,I}{RT}$
9	$t = \frac{s-v}{a}$	**10**	$p = \frac{a-3tq}{3t}$
11	$h = \frac{2A}{b}$	**12**	$B = \frac{2A-hb}{h}$
13	$h = \frac{3V}{r^2}$	**14**	$C = \frac{5F-160}{9}$

Simultaneous equations (pp. 17–23)

1 Substitution (pp. 17–18)

1	(2, 1)	**2**	(3, 2)
3	(-3, 9)	**4**	(9, 1)
5	(7, 8)	**6**	(2, 6)
7	(8, -1)	**8**	(5, 2)
9	(-1, 1)	**10**	(7, -1)

2 Elimination (pp. 19–20)

1	(4, 2)	**2**	(6, -2)
3	(-1, 6)	**4**	(2, 5)
5	(-1, 8)	**6**	(4, 6)
7	(6, 5)	**8**	(4, -2)
9	(7, 1)	**10**	(2, -2)
11	(4, 9)	**12**	(4, -2)

ISBN: 9780170370431

Forming and solving simultaneous equations (pp. 21–23)

1. Jeggings cost \$38 and shorts cost \$28.
2. Ice blocks cost \$1.20 and juicies cost \$0.65.
3. 21 boys and 7 girls.
4. 7 Achieved and 13 Merit grades.
5. Adam has 48 and Chloe has 16.
6. Pens cost \$0.85 and exercise books cost \$1.35.
7. Kowhai trees cost \$28 and lancewoods cost \$7.
8. Peaches cost \$0.75 and apples cost \$0.45.
9. Pies cost \$1.05 and burgers cost \$1.75.
10. T-shirts cost \$35 and singlets cost \$25.
11. They sold 68 packets of seeds and 27 calendars.

Straight lines (pp. 24–61)

Co-ordinates revision (pp. 24–25)

a	(2, 6)	**b**	(6, 4)
c	(-3, 7)	**d**	(2, -3)
e	(-3, -6)	**f**	(0, 7)
g	(8, -9)	**h**	(-7, -4)
i	(4, 10)	**j**	(0, -9)
k	(-6, 8)	**l**	(4, 0)
m	(-9, -10)	**n**	(-5, 0)
o	(-10, 2)	**p**	(0, 0)

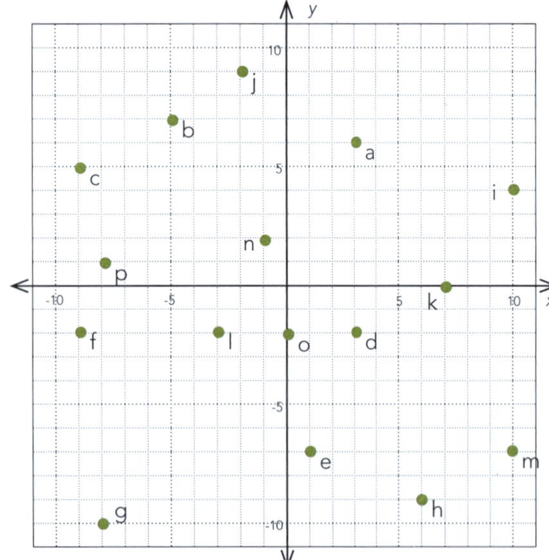

Linear patterns with discrete data (pp. 26–32)

1 **a** $B = 3n + 1$

b

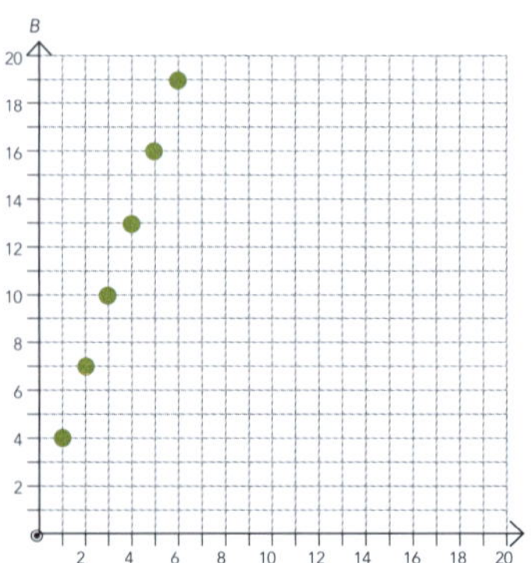

c 30th pattern needs 91 buttons.

d The 90th pattern would need 271 buttons.

e 3 represents how many buttons need to be added for each extra pattern.
1 represents the single button at the centre of every pattern.

2 **a** $B = 2n + 4$

b

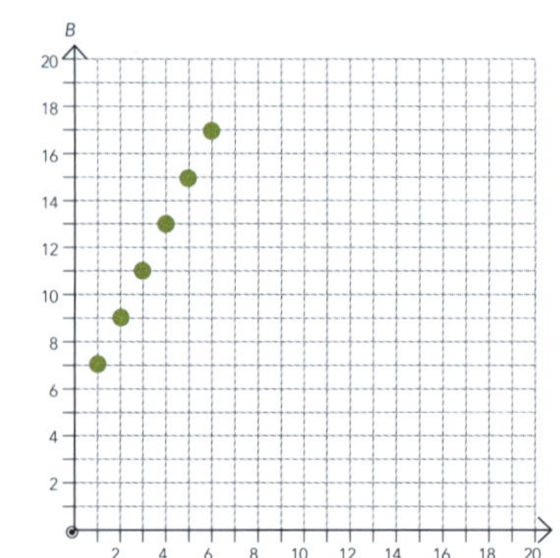

c 30th pattern needs 64 buttons.

d The 80th pattern would need 164 buttons.

e 2 represents the number of buttons that needs to be added for each extra pattern.
4 represents the 4 buttons at the centre of every pattern.

3 **a** $M = 4n + 1$

b

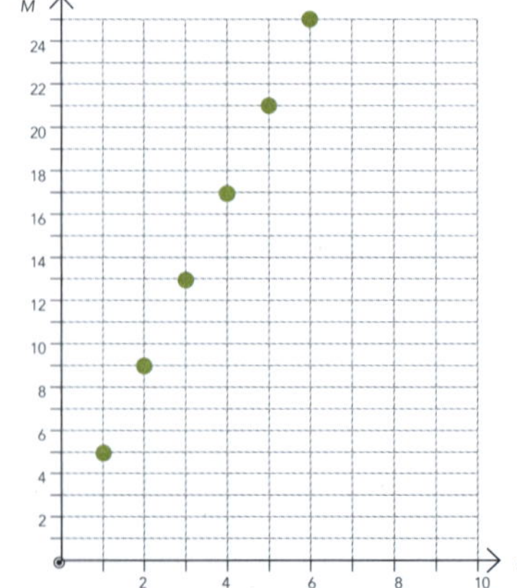

c 30th pattern needs 121 matchsticks.

d The 61st pattern would need 245 matchsticks.

ISBN: 9780170370431

e The 4 represents the number of matchsticks that needs to be added for each extra pattern. The 1 represents the single upright matchstick on the left that every pattern has.

4 **a** $D = 10s + 40$

b

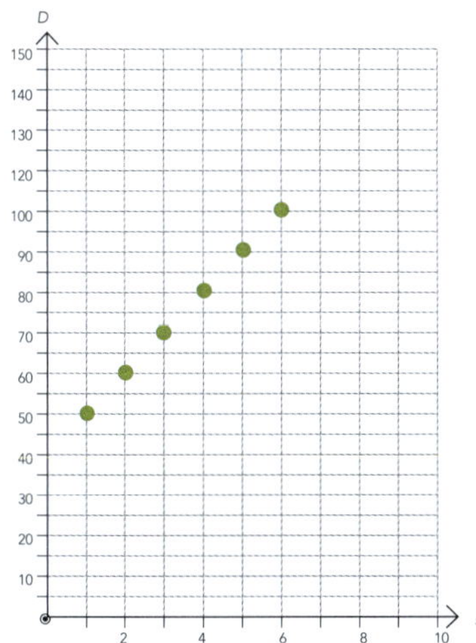

c If he passes 13 standards with Merit or Excellence, he earns \$170.

d If he earned \$210, he passed 17 standards with Merit or Excellence.

e The 10 represents how much he earns for each standard he passed with Merit or Excellence. The 40 represents the \$40 he gets for passing Level 1.

5 **a** $G = 4T + 2$

b

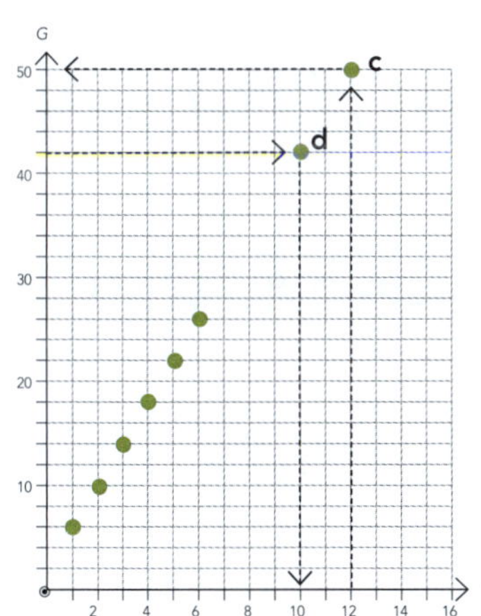

c 50 guests (see graph).
Its point is in line with the points from **b**.

d 10 tables (see graph).
Its point is in line with the points from **b**.

e The 4 represents the number of extra people that can be added with each extra table. The 2 represents the two people who sit at each end, regardless of how many tables there are.

6 **a** $D = -15W + 390$

b

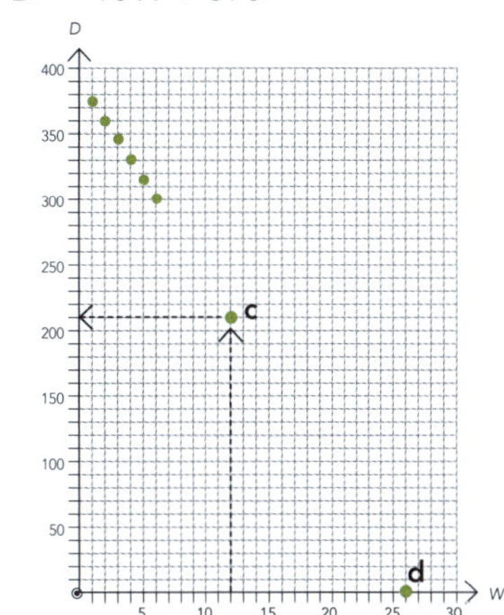

c He owes \$210 (see graph).
Its point is in line with the points from **b**.

d At the end of week 26.
Its point is in line with the points from **b**.

e The -\$15 represents how much he pays each week. The \$390 represents the amount he borrowed from his mother.

The gradient of a line (pp. 33–35)

1 **a** $\frac{4}{2} = 2$ **b** $\frac{3}{2}$

c $\frac{2}{5}$ **d** $\frac{0}{5} = 0$

e $-\frac{2}{4} = -\frac{1}{2}$ **f** $-\frac{5}{2}$

g $\frac{2}{2} = 1$ **h** $\frac{3}{0}$ = undefined

i $\frac{1}{3}$ **j** $-\frac{1}{7}$

2

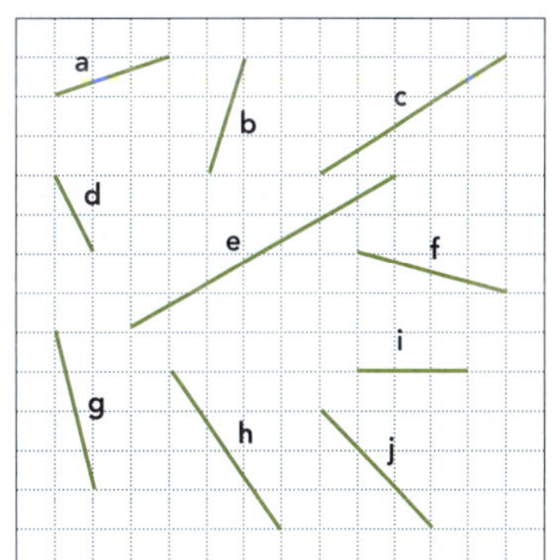

Gradients with different scales (pp. 36–39)

1 Drawing gradients (pp. 36–37)

1

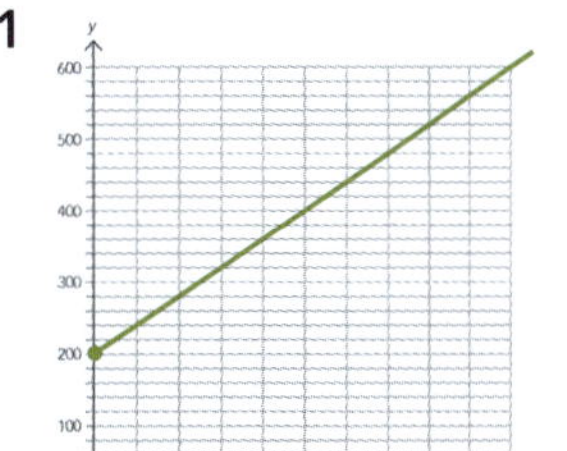

2

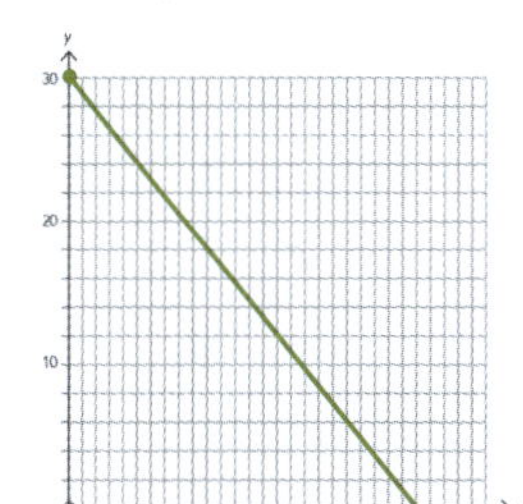

3

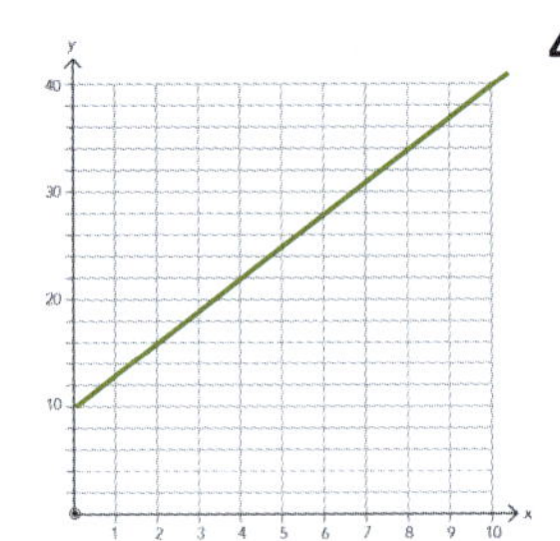

4

5

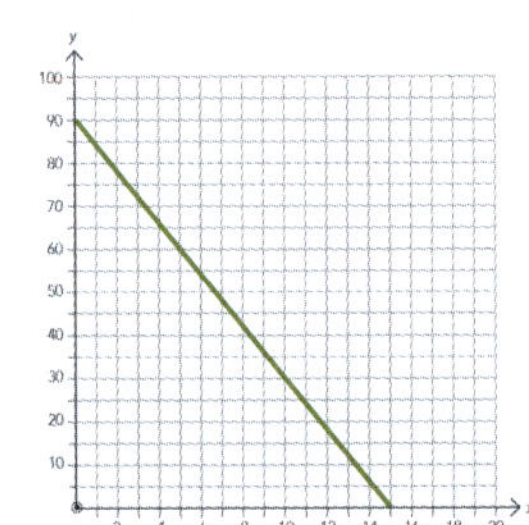

6

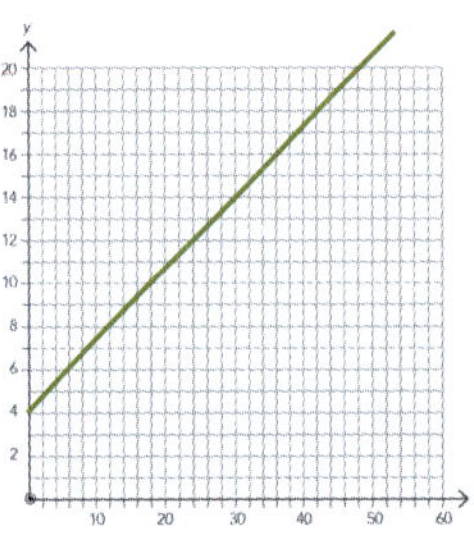

7

8

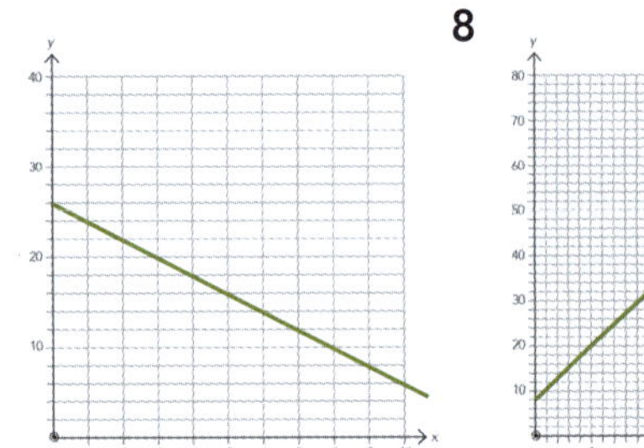

2 Calculating gradients (pp. 38–39)

1	$m = 50$	**2**	$m = \frac{1}{2}$
3	$m = -\frac{1}{2}$	**4**	$m = -5$
5	$m = -4$	**6**	$m = \frac{2}{5}$
7	$m = -\frac{1}{4}$	**8**	$m = 8$

Drawing straight lines — continuous data (pp. 40–50)

1 Plotting points using the equation (pp. 40–42)

1 & 2

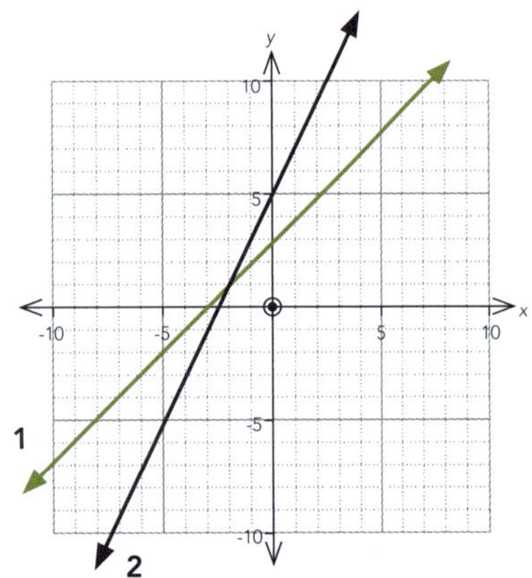

3 & 4

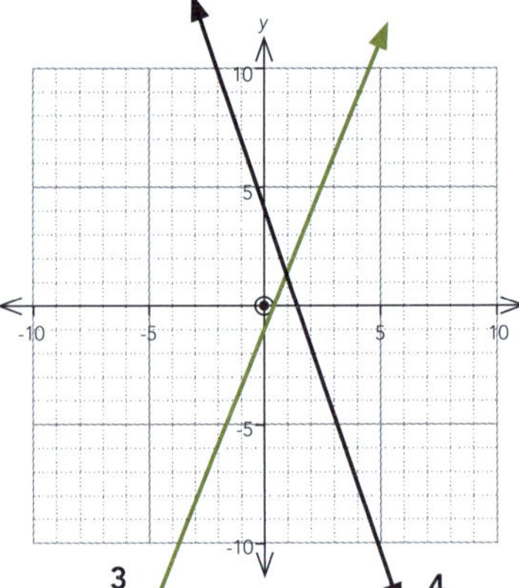

5 & 6

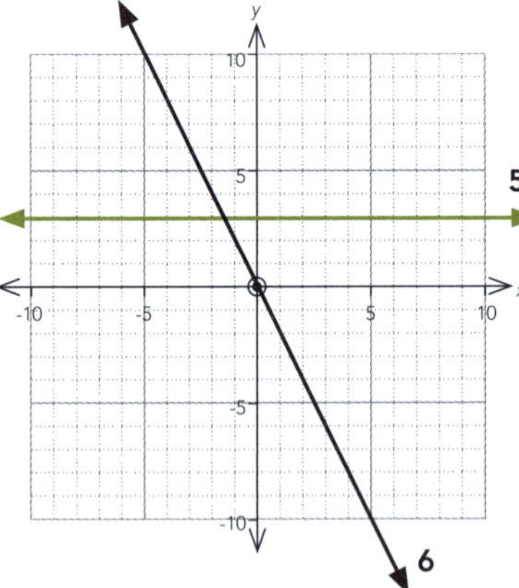

2 Using the y-intercept and the gradient (pp. 43–45)

1 & 2

3 & 4

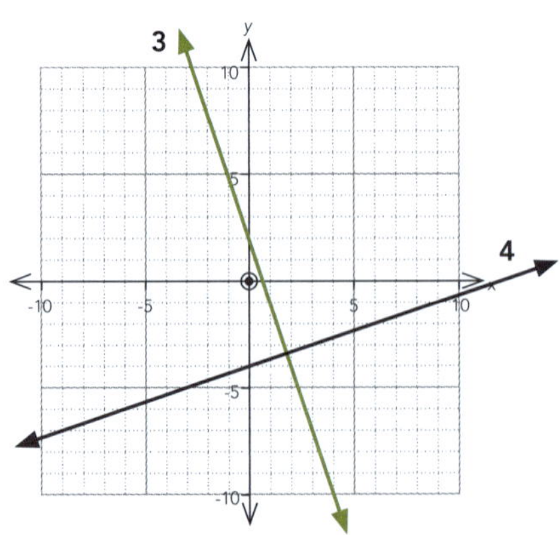

 ISBN: 9780170370431

5 & 6

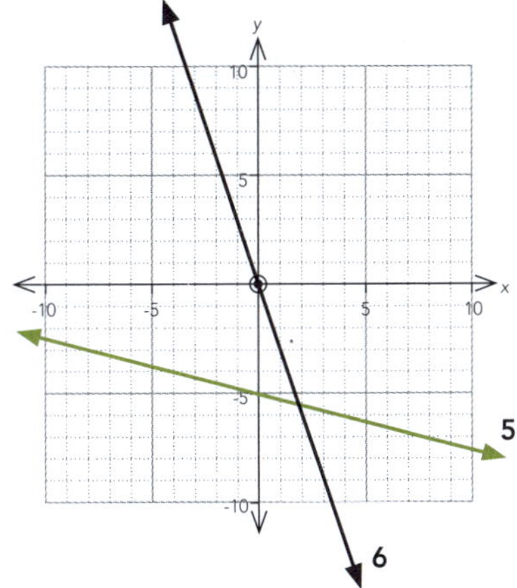

7 & 8

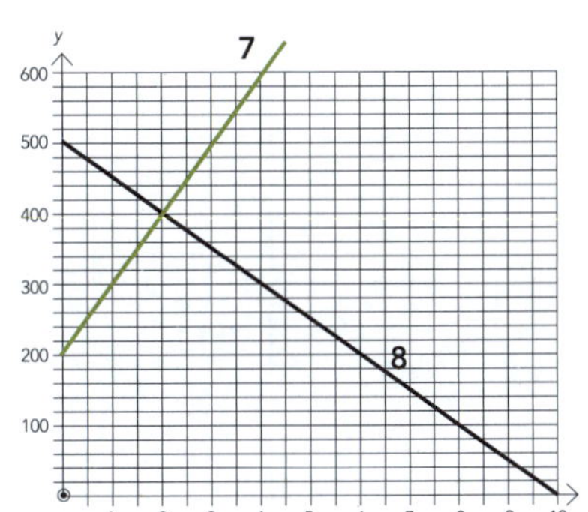

9 & 10

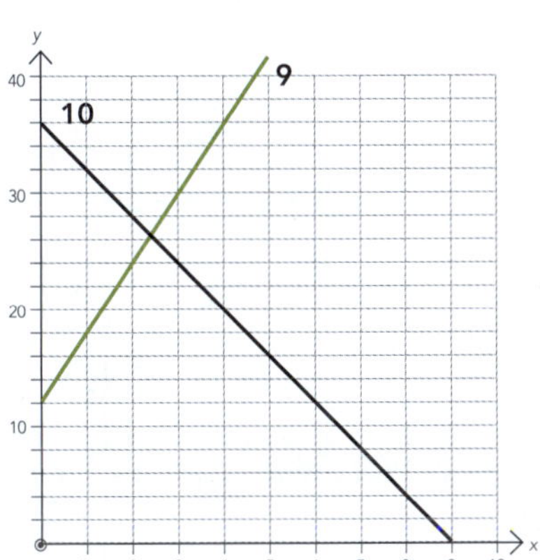

11 & 12

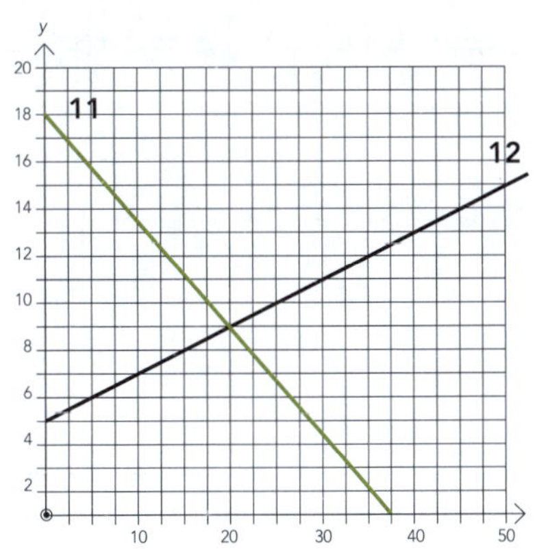

3 Using the *x*- and *y*-intercepts (pp. 46–48)

1 & 2

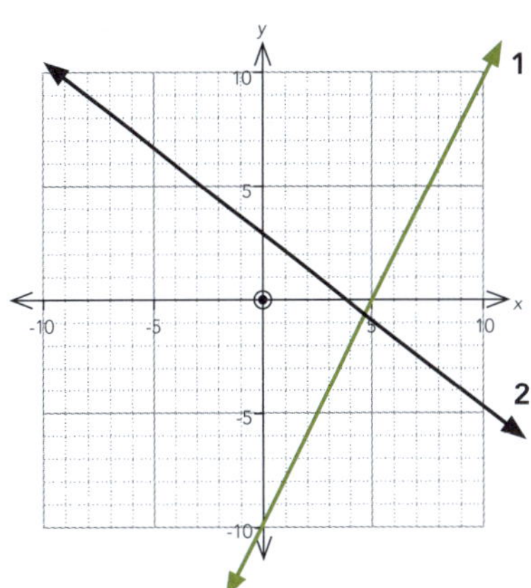

3 & 4

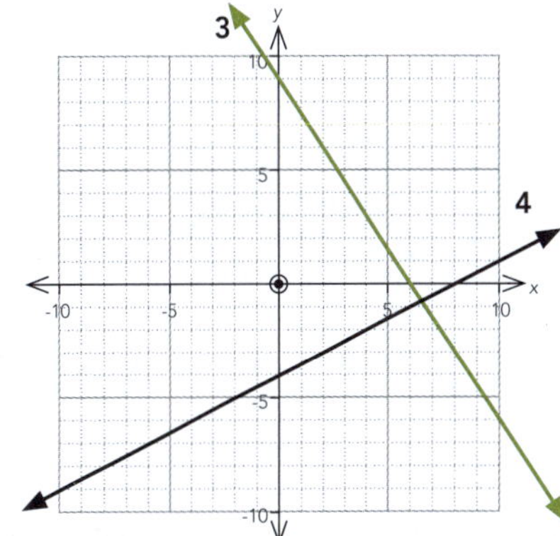

5 & 6

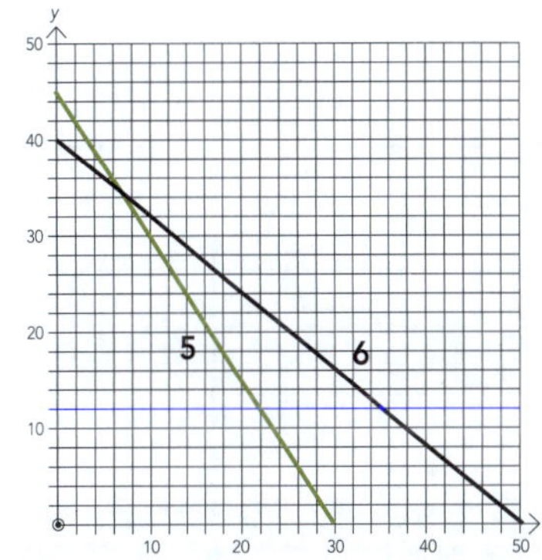

7 & 8

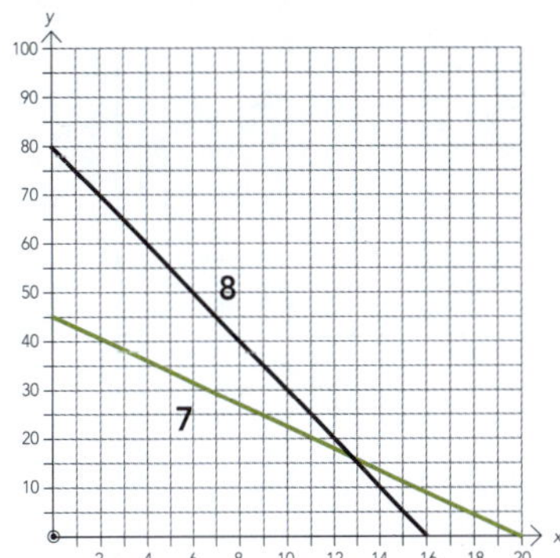

ISBN: 9780170370431

9 & 10

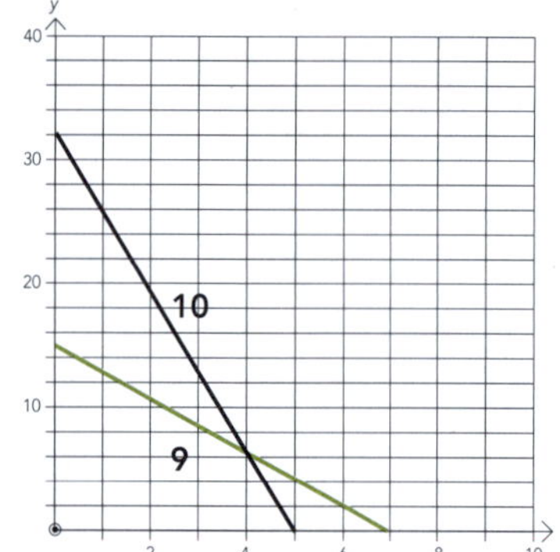

11 & 12

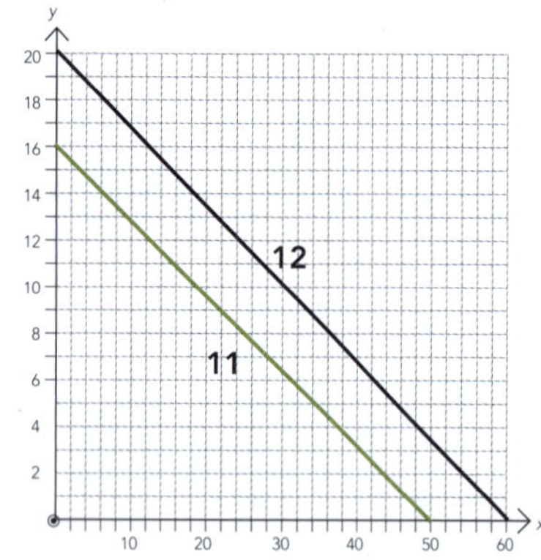

4 Horizontal and vertical lines (pp. 49–50)

1

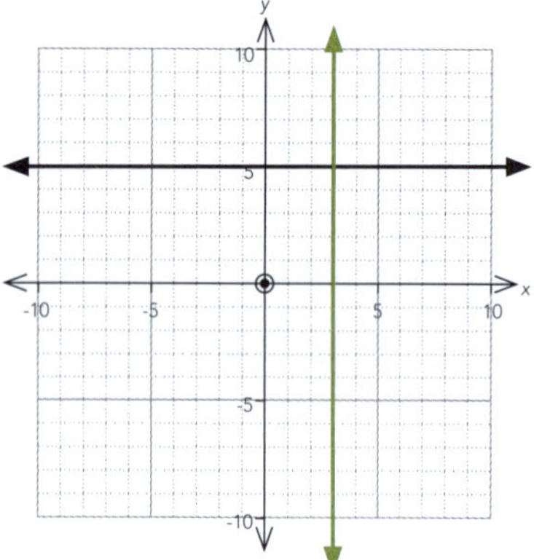

2

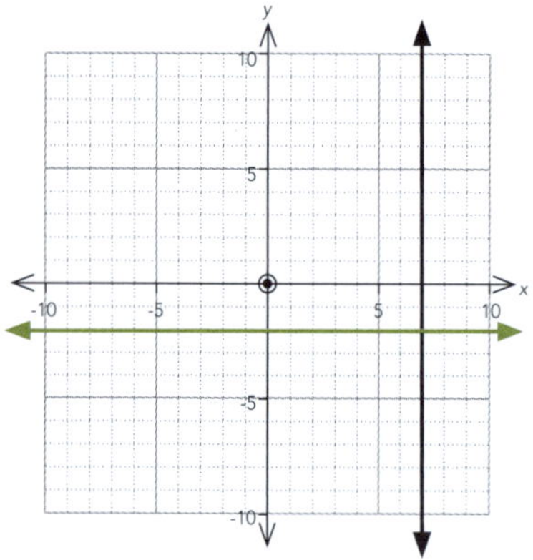

3

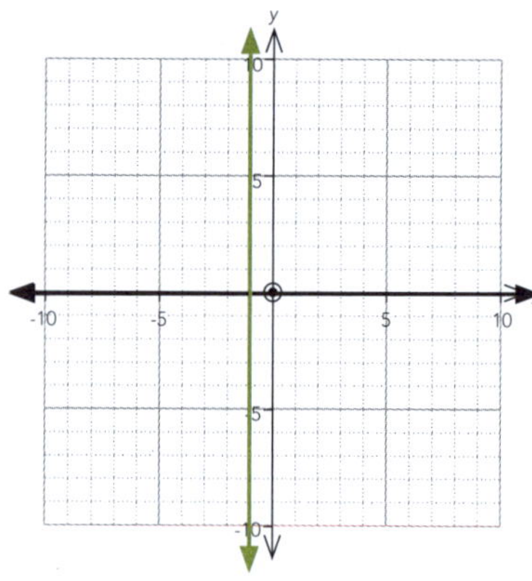

4

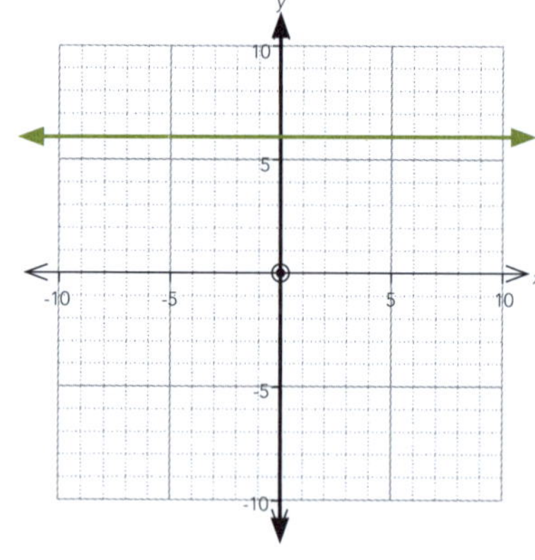

5

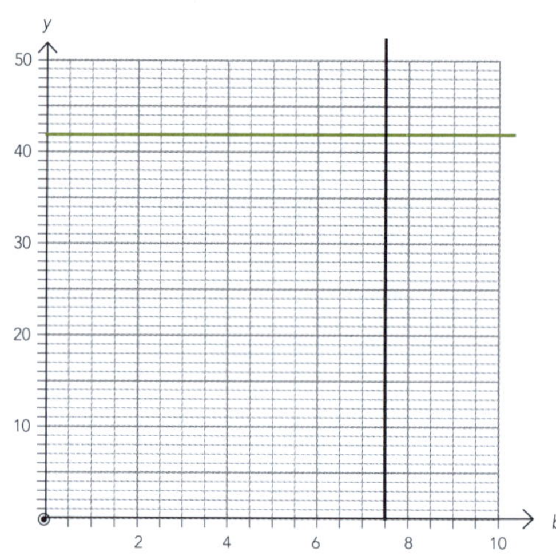

6

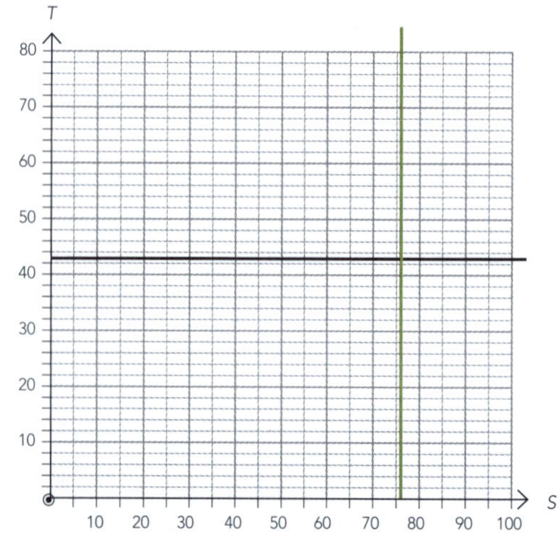

 ISBN: 9780170370431

Writing equations from graphs (pp. 51–57)

1 Using the gradient and the y-intercept (pp. 51–54)

1	$y = 2x + 3$	**2**	$y = -x - 3$
3	$y = -4x + 5$	**4**	$y = \frac{1}{2}x - 3$
5	$y = \frac{1}{2}x + 3$	**6**	$y = -6x + 90$
7	$y = -5x + 85$	**8**	$y = -\frac{1}{2}x + 38$
9	$y = \frac{1}{4}x + 12$	**10**	$y = -11x + 66$
11	$y = \frac{2}{5}x + 8$	**12**	$y = 10x + 16$

2 Using two points (pp. 55–57)

1	$y = 2x + 3$	**2**	$y = \frac{1}{2}x + 7$
3	$y = x - 4$	**4**	$y = 2x + 5$
5	$y = -2x + 3$	**6**	$y = -x + 6$
7	$y = -\frac{1}{2}x + 1$	**8**	$y = -4x - 3$
9	$y = -\frac{1}{3}x + 8$	**10**	$y = \frac{2}{3}x - 5$
11	$y = 11x - 12$	**12**	$y = -12x + 43$
13	$y = -2x - 17$	**14**	$y = -\frac{2}{5}x + 13$
15	$y = \frac{3}{4}x - 14$	**16**	$y = -\frac{1}{4}x - 13$

Applications (pp. 58–61)

1 **a** \$120

It is where the line meets the *S*-axis.

b \$30

Gradient = $\frac{480}{16} = 30$

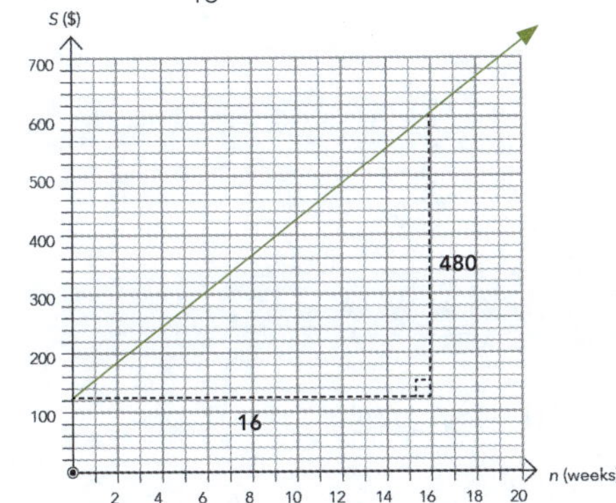

c $S = 30n + 120$

d 18 weeks

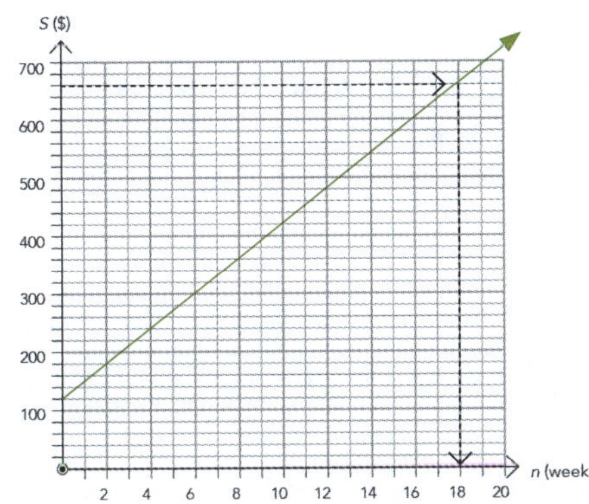

e

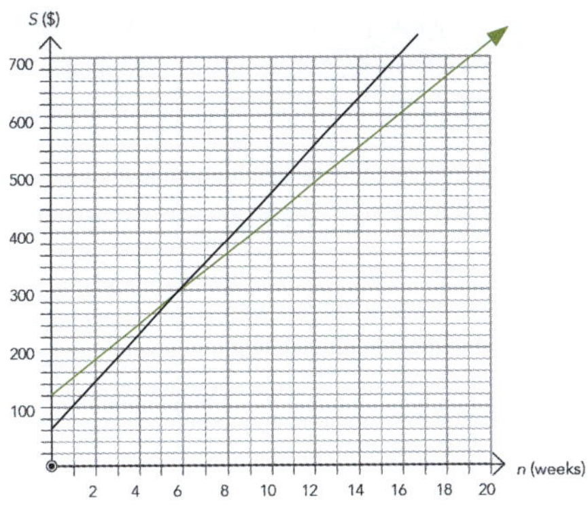

f $S = 40n + 60$

g After 6 weeks. This is the point where the lines cross each other. They have saved both \$300.

2 **a** \$60

This is where the line meets the *C*-axis.

b \$0.80

Gradient = $\frac{240}{300} = 0.8$

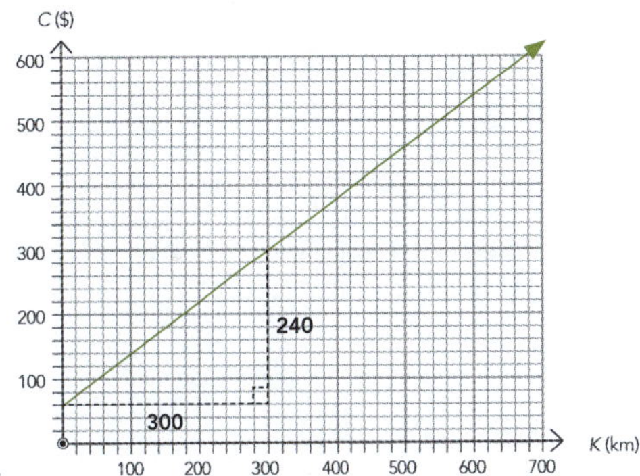

c $C = 0.8K + 60$

d \$220

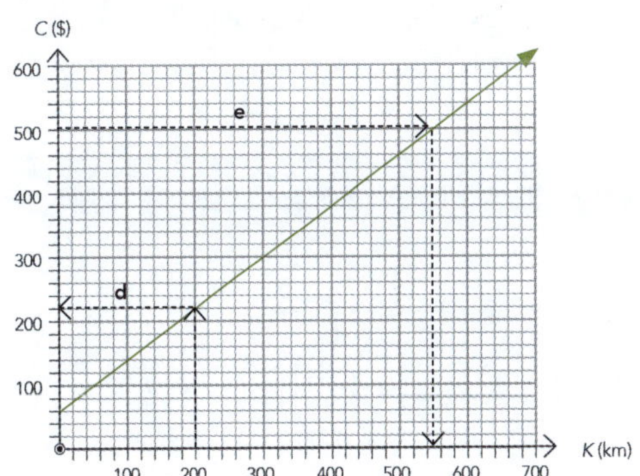

e 550 km

f

g 300 km.
This is the point where the lines cross.
\$300

h $C = 0.6K + 120$

i If they think they will drive less than 300 km, then they should use Vicky's Vans. If they think they will drive more than 300 km, they should use Rip-off Rentals.

3 **a** \$480. This is the P-intercept.

b \$15
Gradient $= -\frac{480}{32} = -15$

c $P = -15W + 480$

d \$120

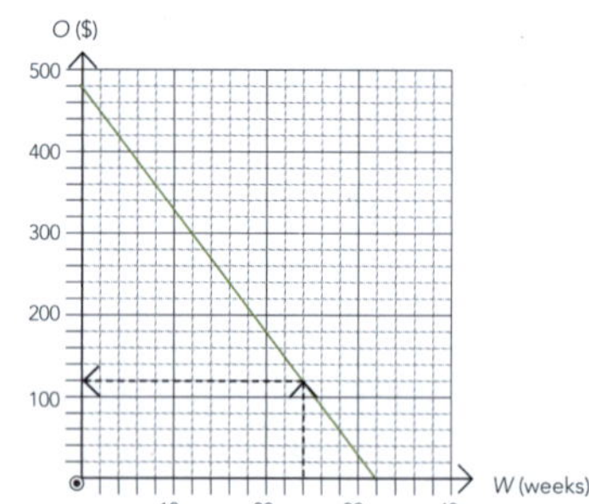

e 32 weeks
This is the point where the line cuts the W-axis.

$$P = -15W + 480$$
$$P = 0 \Rightarrow 0 = -15W + 480$$
$$15W = 480$$
$$W = 32$$

f

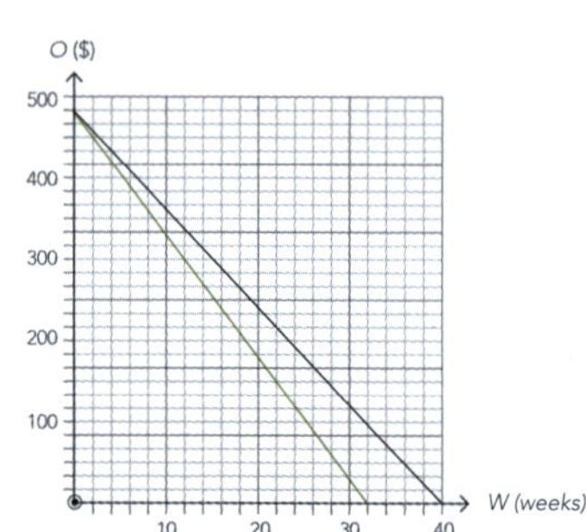

g 8 weeks

h $P = -12W + 480$

4 **a** \$3400
It is where the line meets the S-axis.

b \$250
Gradient $= -\frac{2000}{8} = -\$250$

c $S = -250W + 3400$

d \$1400

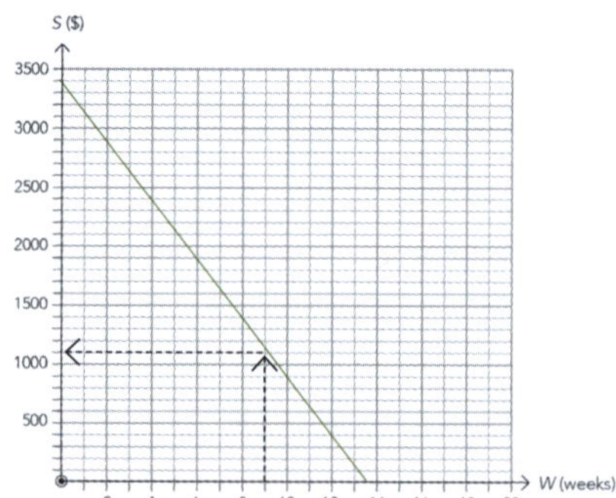

e At the end of week 12.

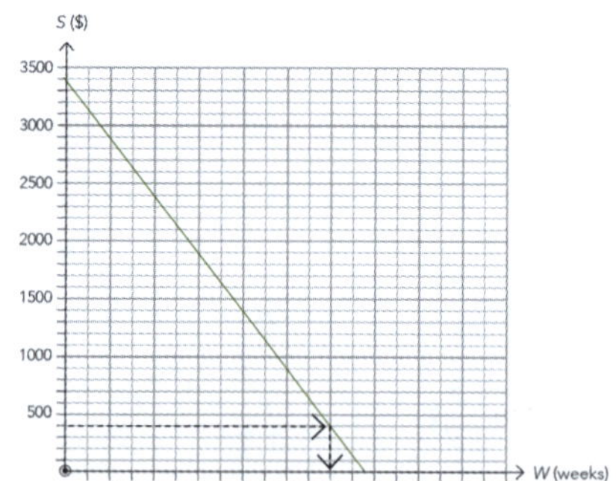

f

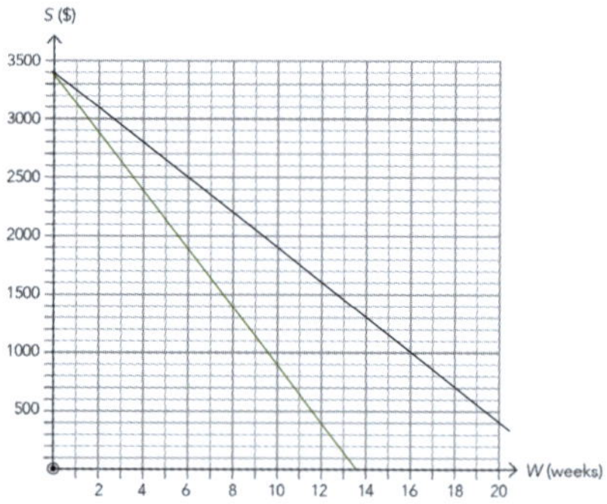

g $S = -150W + 3400$

h Lasts 22.7 weeks.

i \$1000

Inequations (pp. 62–73)

1 $x \geq 3$
2 $x < 12$
3 $x \geq -2$
4 $x < -5$
5 $x < 2$
6 $x \leq -43$
7 $x < -1.4$
8 $x < -7$

Forming and solving inequations (p. 63)

1 $17 - 5x > 2$
$x < 3$

2 $2.5x + x + 2.5x + x \leq 112$
$x \leq 16$
Maximum dimensions are 16 cm by 40 cm.

3 $\frac{3}{4}A + A + (\frac{3}{4}A - 4) \leq 66$
Alannah's: less than \$28.
Meg's: less than \$21.
Suzie's: less than \$17.

ISBN: 9780170370431

4 $48 + 7.5x \leq 95$
$x \leq 6.27 \Rightarrow$ she can invite 5 people. (Don't forget that she needs a ticket too.)

5 $p \geq 6.50 + \frac{3}{5}s$
Pete has at least \$14.60.

Graphing inequations (pp. 64–67)

1 $x > 6$
2 $y \leq 2$
3 $y \geq x$
4 $y < -x + 7$
5 $y > \frac{1}{2}x + 4$
6 $y \geq -2x + 10$
7 $5 < 4p + 5$
8 $c \geq -0.5f + 45$
9 $s < 4p + 5$
10

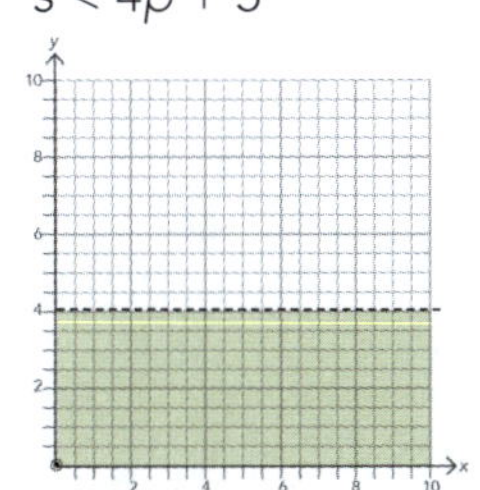

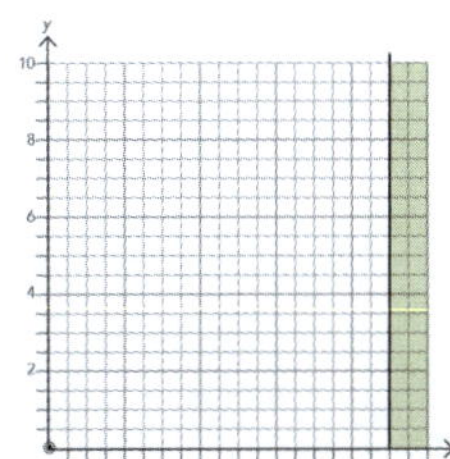

11

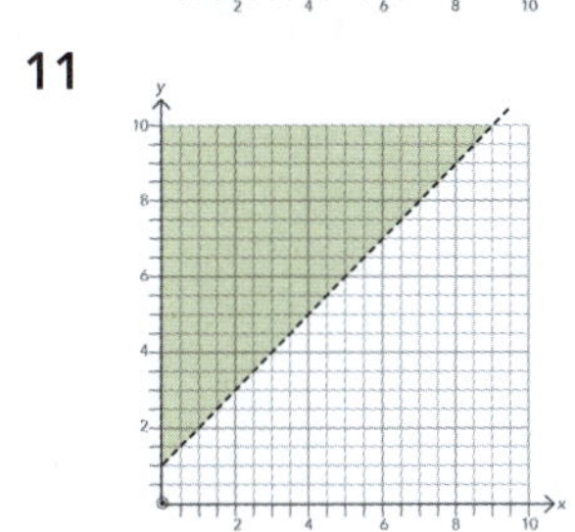

12

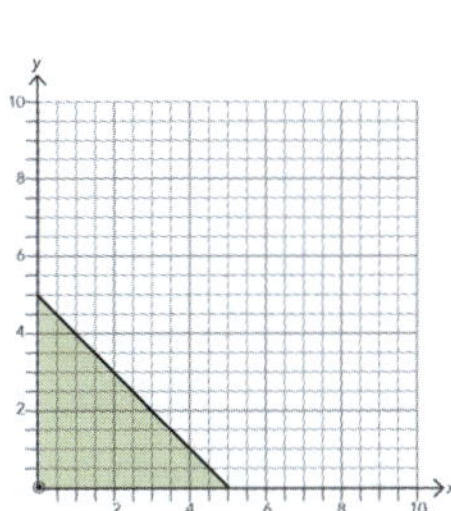

13

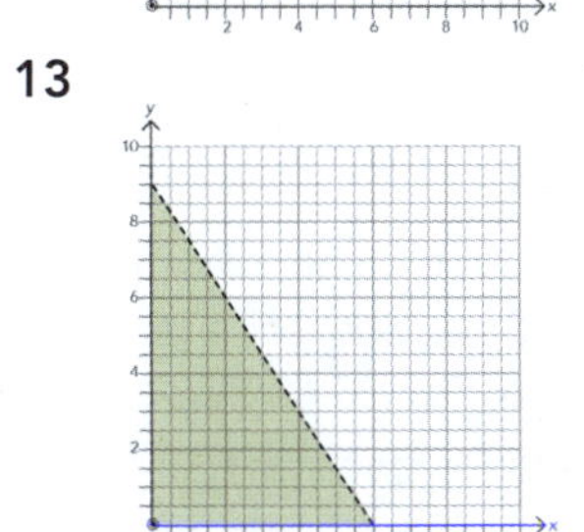

14

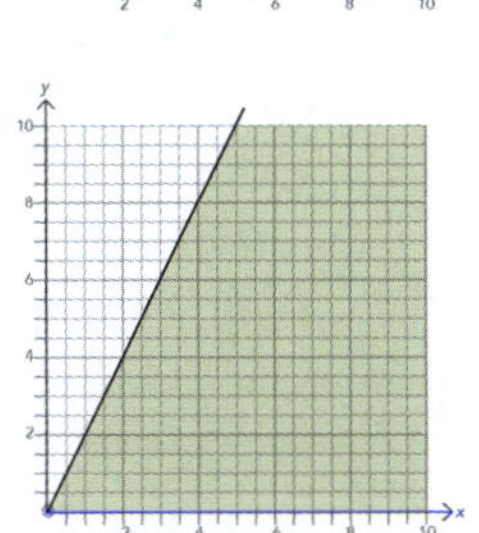

Combining several inequations (pp. 68–73)

1 $s > 3a$
$s = a + 8$

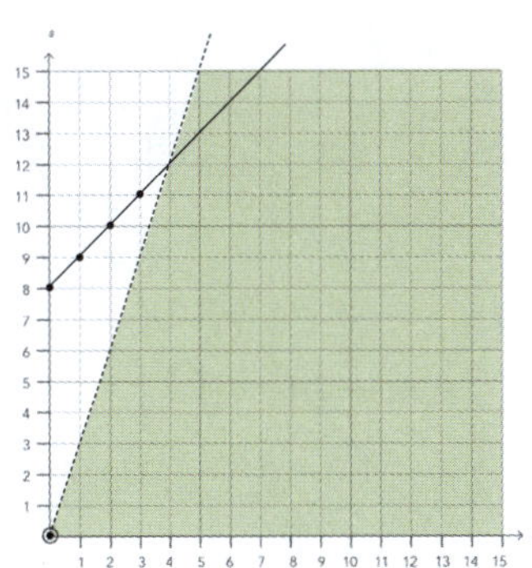

Minimum: Alexander has none and Sebastien has 8. Maximum: Alexander has 3 and Sebastien has 11.

2 $f \geq r + 5$
$2f = r + 16$

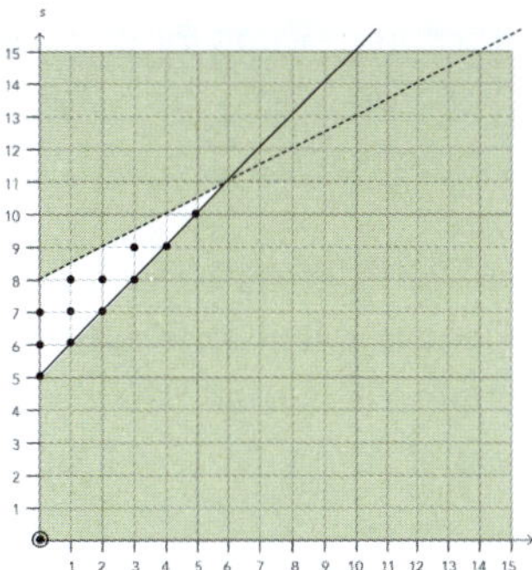

Maximum: Rosa has 5 and Florrie has 10.

3 Bob's Bounce: $C = 80 + 20h$
Craig's Castles: $C \geq 40h$

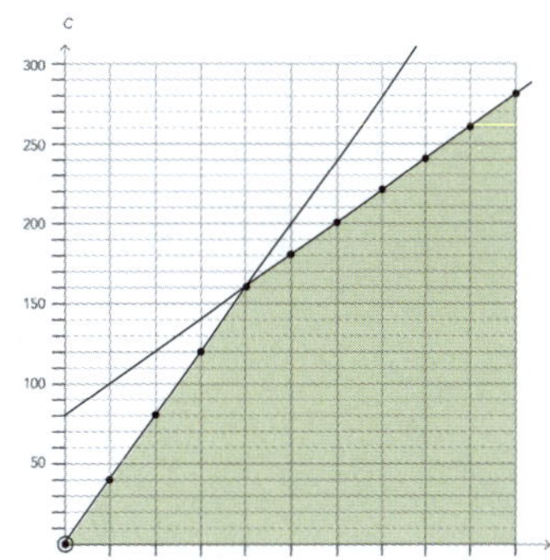

For 1, 2 or 3 hours: Craig's Castles are cheaper, provided there is no damage.
Example: Three hours' hire costs a minimum of \$120 from Craig's Castles, while Bob's Bounce would cost \$140.
For 4 hours: Both cost \$160, but if there was damage, then Craig's Castles would charge more.
For 5 to 10 hours: Bob's Bounce is definitely cheapest.
Example: Seven hours' hire would cost \$\$220 from Bob's Bounce, whereas Craig's Castles would cost a minimum of \$280.

4 Chrissie's Canvas: $C = 20h + 100$
Tom's Tents: $C < 200$

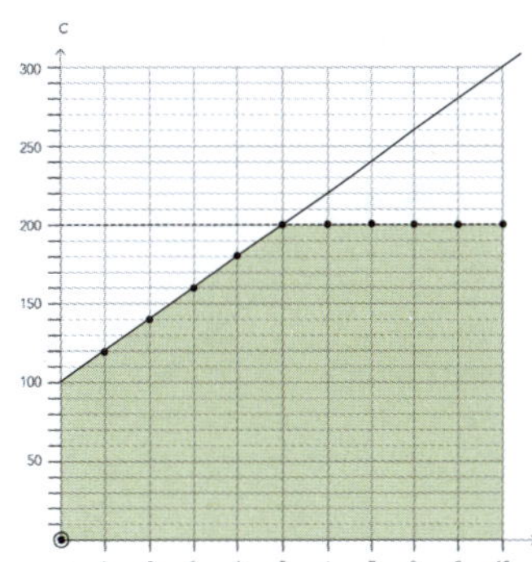

For 1–4 hours: Chrissie's Canvas is probably the cheapest.

ISBN: 9780170370431

Example: For three hours' hire, Chrissie's Canvas will charge \$160. Tom's Tents could charge anything up to \$200, so they could be cheaper.
For 5 hours: Tom's Tents is definitely cheaper. Chrissie's Canvas will charge \$200, but Tom's Tents will charge less than \$200.
For 6–10 hours: Tom's Tents will definitely be cheaper.
Example: For eight hours' hire, Tom's Tents will charge less than \$200, but Chrissie's Canvas will charge \$260.

Piecewise functions (pp. 74–78)

1

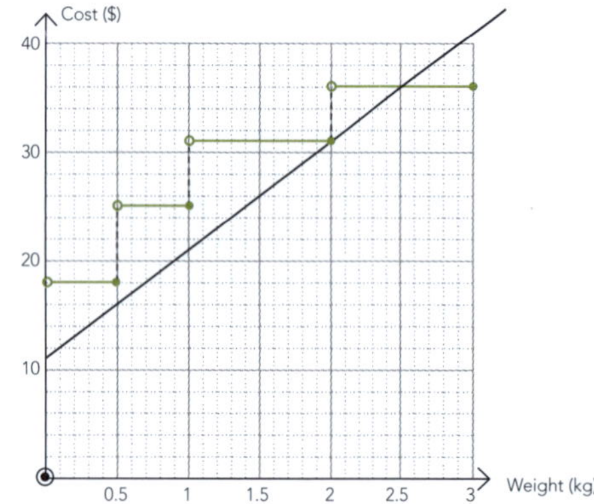

For weights less than 2.5 kg it is generally cheaper to use the courier. Example:
A 1.5 kg parcel costs \$31 if posted, but only \$26 by courier.
There is one exception: a parcel weighing exactly 2 kg costs \$31 either way.
A parcel weighing exactly 2.5 kg costs \$36 either way.
Above 2.5 kg it is cheaper to post it. Example:
A 2.8 kg parcel costs \$39 by courier, but only \$36 if posted.

2

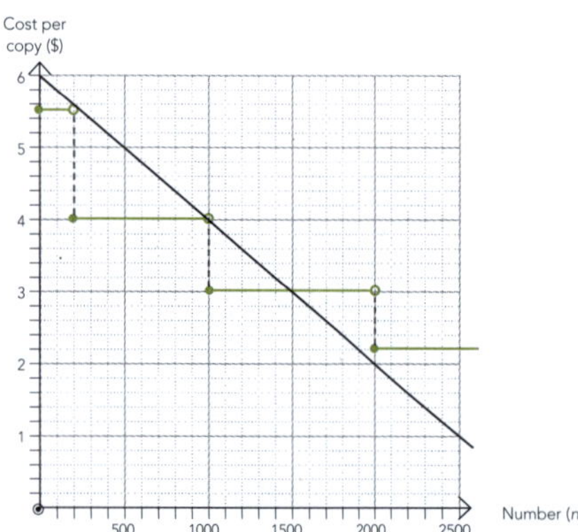

If you were printing fewer than 1500 magazines, Pete's Printing would generally be cheaper.
Example: 900 magazines would cost \$4.00 each from Pete's Printing, but \$4.20 from Maria's Mags.
For exactly 1500 magazines, they also charge the same amount — \$3.00 each.
For more than 1500 magazines, it is always cheapest to use Maria's Mags. Example: 2000 magazines would cost \$2.20 from Pete's Printing, but only \$2.00 from Maria's Mags.

3 **a**

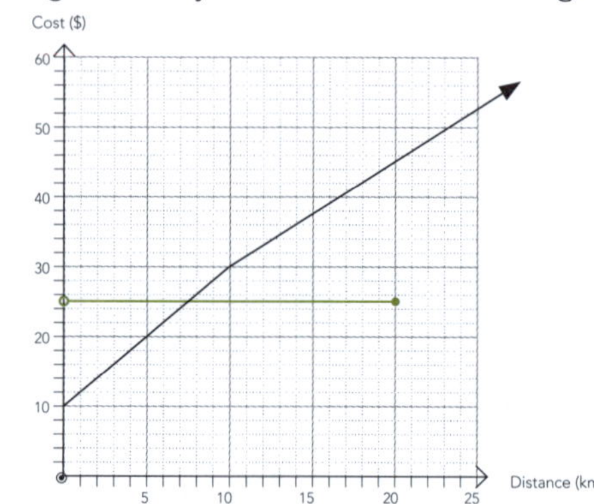

b $c = 2d + 10;\ d \leq 10$
$c = 1.5d + 15;\ d > 10$

c 9 km costs \$28.
18 km costs \$42.

d $c = 25;\ 0 < d \leq 20$

e For distances of less than 7.5 km, Tiger Taxis are definitely cheaper. Example: For 6 km, Tiger Taxis costs \$22, but Claude's Cabs would be \$25.
For exactly 7.5 km, both charge \$25.
For distances over 7.5 km but less than 20 km, Claude's Cabs are cheapest. Example: For 20 km, Claude's Cabs charge \$25 and Tiger Taxis charge \$45. For distance of more than 20 km, customers have no choice, they can use only Tiger Taxis.

4 **a**

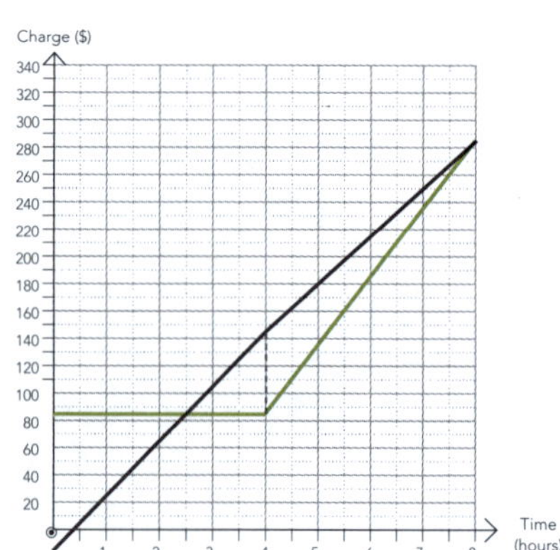

b $c = 40t + 30;\ 0 < t \leq 4$
$c = 35t + 50;\ t > 4$

c 3 hours' work costs \$150.
8 hours' work costs \$330.

  ISBN: 9780170370431

d $c = 130;\ 0 < t \le 4$
$c = 50t - 70;\ t > 4$

e For times of less than 2.5 hours, Penny is definitely cheaper. Example: For 2 hours, Penny charges \$110, but No Drips would charge \$130. For exactly 2.5 hours, both charge \$130. For times over 2.5 hours but less than 8 hours, No Drips are cheapest. Example: For 5 hours, No Drips charge \$180 and Penny charges \$225.
For 8 hours' work they both charge \$325.

Practice tasks (pp. 79–84)

Practice task one (pp. 79–80)

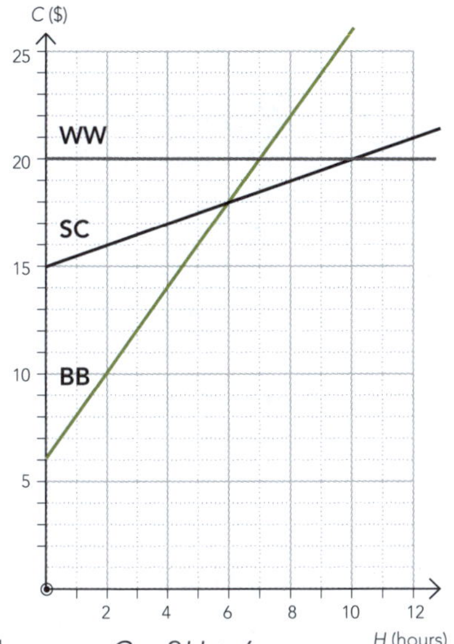

Ben's Bikes: $C = 2H + 6$
Sally's Cycles: $C = 1.5H + 15$
Will's Wheels: $C = 20$

Hemi: Should definitely use Ben's Bikes because it will cost him only \$10, whereas Sally's Cycles would charge \$16 and Will's Wheels would charge \$20.
Cleo: Could use either Ben's Bikes or Sally's Cycles because both would cost \$18. Will's Wheels would charge \$20.
Didier: Should use Sally's Cycles because she would charge \$19.50. Ben's Bikes would charge \$24 and Will's Wheels would charge \$20. However, if he thought that he might want the bike for a little longer than 9 hours, he would be better to use Will's Wheels, who will only charge him 50c more for the whole day.

Sally's Cycles should be used for a person who wants a bike for between 6 and 10 hours. Any less, and Ben's bikes are cheaper, and for any longer it is best to use Will's Wheels.

If Ben wants to be cheapest he could:

1. Charge a fixed charge of \$15 and \$0.40 per hour (shown as dashed line). This will be cheaper for up to 12 hours. For 12 hours, Ben would charge \$19.80, which is cheaper than Will's Wheels (\$20). For 13 hours or more, Will's Wheels would be cheaper at \$20 because 13 hours at Ben's cheaper rate would be \$20.20.
2. Charge a fixed charge of \$10 and \$0.75 per hour (shown as dotted line). This will be cheaper for up to 13 hours. For 13 hours, Ben would charge \$19.75, which is cheaper than Will's Wheels (\$20). For 14 hours or more, Will's Wheels would be cheaper at \$20 because 14 hours at Ben's cheaper rate would be \$20.50.

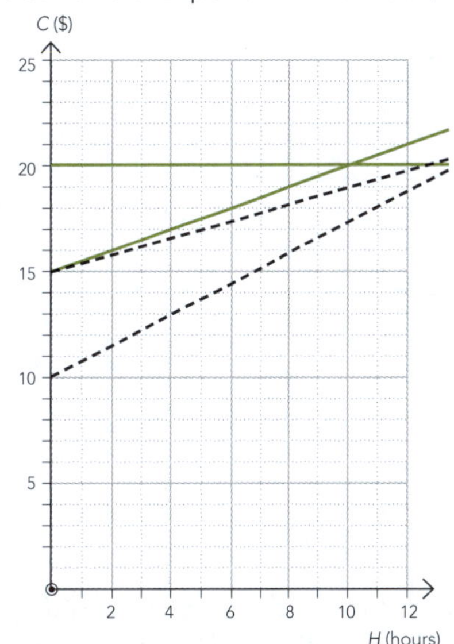

Practice task two (pp. 81–82)

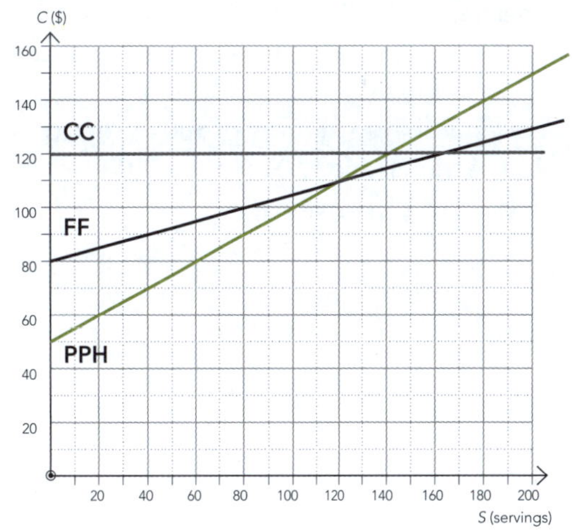

Felicity's Floss: $C = 0.25S + 80$
Pita's Party Hire: $C = 0.50S + 50$
Colin's Candy: $C = 120$

For fewer than 120 servings:
They should use Pita's Party Hire because theirs is the lowest line on the graph.

For example: 100 servings — Pita would charge \$100, Felicity would charge \$105 and Colin would charge \$120.

For between 120 and 160 servings:
They should use Felicity's Floss because hers is the lowest line on the graph. For example: 140 servings — Felicity would charge \$115, and both the other two companies would charge \$120.

For more than 160 servings:
They should use Colin's Candy because his is the lowest line on the graph. For example: 200 servings — Colin would charge \$120, Felicity would charge \$130 and Pita would charge \$150.

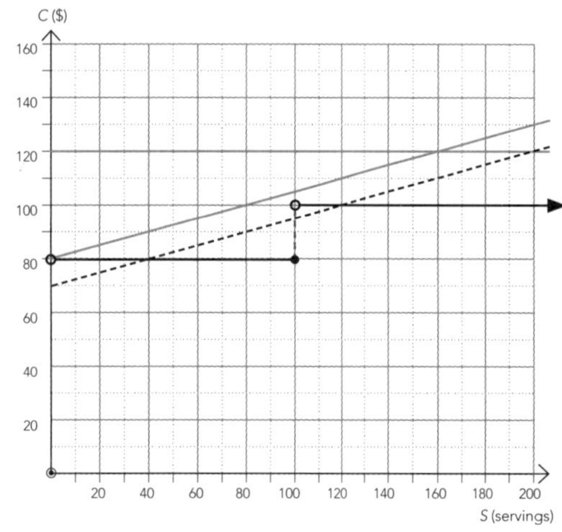

If Pita wants to be cheapest he could:

1 Have a fixed charge of \$70 and then charge \$0.25 per serving. If he did this, his total charge would always be \$10 less than Felicity's because, while his charge per serving is the same, his fixed charge is \$10 less.

2 He could have flat rates of \$80 for up to 100 servings, and \$100 for over 100 servings and up to 200 servings.

Practice task three (pp. 83–84)

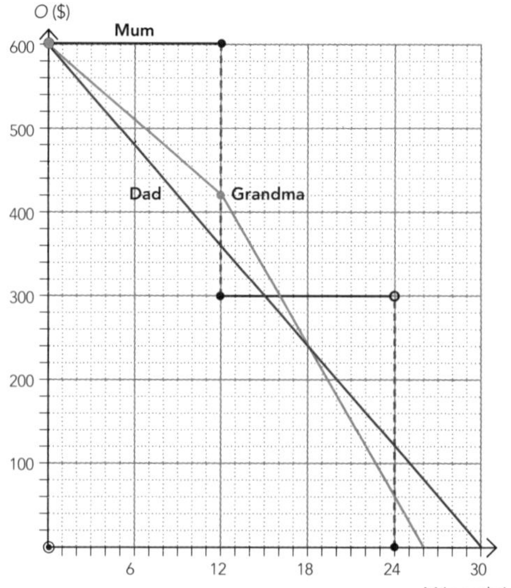

Mum: $O = 600$; $M < 12$
$O = 300$; $12 \leq M < 24$
$O = 0$; $M \geq 24$
Dad: $O = 600 - 20M$
Grandma: $O = 600 - 15M$; $M < 12$
$O = 840 - 30M$; $M \geq 12$

How long she will take to repay:
Mum: 24 months
Dad: 30 months
Grandma: 26 months

After 12 months she owes:
Mum: \$300
Dad: \$360
Grandma: \$420

After 18 months she owes:
Mum: \$300
Dad: \$240
Grandma: \$240

After 24 months she owes:
Mum: \$0
Dad: \$120
Grandma: \$60

Grandpa's repayment schemes:

1 She could repay him at \$25 per month ($O = 500 - 25M$) and she would pay the money back in 20 weeks, which is faster than she would pay her dad, her grandma or her mum.

2 She could repay him at \$20 per month ($O = 500 - 20M$) and she would pay the money back in 25 weeks, which is faster than she would pay her dad or grandma, but slower than she would pay her mum.

ISBN: 9780170370431